Integrative Statistics for Behavioral Science

Renee R. Ha
James C. Ha
University of Washington
Seattle, Washington

McGraw-Hill Custom Publishing

Boston Burr Ridge, IL Dubuque, IA Madison, WI New York
San Francisco St. Louis Bangkok Bogotá Caracas Kuala Lumpur
Lisbon London Madrid Mexico City Milan Montreal New Delhi
Santiago Seoul Singapore Sydney Taipei Toronto

The McGraw-Hill Companies

Integrative Statistics for Behavioral Science

McGraw-Hill's Custom Publishing series consists of products that are produced from camera-ready copy. Peer review, class testing, and accuracy are primarily the responsibility of the author(s).

1 2 3 4 5 6 7 8 9 0 QSR QSR 0 9 8 7 6 5 4

ISBN 0-07-312239-4

Editor: Julie Kehrwald
Production Editor: Nina Meyer
Printer/Binder: Quebecor World

Integrative Statistics for Behavioral Science

Brief Table of Contents

Integrative Statistics for Behavioral Science

Detailed Table of Contents

Chapter 1
Introduction to Statistics

What are You Getting Into?

This book addresses the topic of statistics, but exactly what do we mean by statistics? Statistics is a branch of applied mathematics that involves the collection and interpretation of quantitative data, and the use of probability theory to estimate population parameters. In other words, statistics is a field of mathematical study that addresses how data that is collected in various studies (such as the number of accidents at a particular intersection) can be analyzed to determine patterns in the data, and to evaluate those patterns based on the likelihood of that pattern occurring by chance if you were to collect all of the data that would be relevant to your study. Patterns in the data that are considered to be unlikely occurrences if chance alone is operating can then be implicated as occurring for reasons other than chance (such as poor visibility in the intersection).

What is Our Approach to Statistics?

This book is intended for those of us who use statistics as a tool to understand the data that we collect in our research or to understand the data collected by others in their research. Thus, although we will not delve too deeply into the theoretical basis behind the mathematics that we teach, you will definitely be using some of your skills in mathematics in this course. It would be a good idea for you to review some basic college mathematics textbooks if you feel you have difficulty recalling things like the proper order that you follow when you have multiple operations in a single problem (e.g. always do what is in the parentheses first). There will be statistical tests that you will learn to calculate by hand. However, while we feel that requiring some calculations to be done by hand is important in teaching the underlying basis of statistical interpretation, this can be done for a limited number of tests while still accomplishing the conceptual goals of the course.

In addition, we will emphasize how the steps you take in calculating and evaluating a statistical test by hand compare to interpreting a print-out of the same statistical test in two different computer software packages. Statistical terminology varies among textbooks and software packages, so it is critical that you are trained to understand the conceptual basis to the labels. That will be easier for you if you have been exposed to more than one style of output. Truthfully, very few scientists calculate statistical tests by hand, and when they do, it is only for the simple tests. Thus, we will provide examples in the appropriate chapters that demonstrate the step-by-step calculations done by hand and then show the output from the same data in Microsoft's Excel and SPSS Inc.'s Statistical Package for the Social Sciences (SPSS). We chose these software packages because they are much different from one another, but are commonly used, and SPSS has a student version. We will highlight the output that matches the numbers that we calculated by hand, and we will define all of the output since there are more numbers provided by computer software packages than you normally use when you hand calculate the same tests. These additional numbers have proven confusing to students when they get to their laboratory course if they have only been trained in the hand calculation techniques.

Beginning in unit two, you will be introduced to a flowchart that will guide you in choosing the appropriate statistical test for a given data set and experimental design, and that flowchart will be built up chapter by chapter as you learn more tests. We feel that choosing the appropriate test is the most important concept in undergraduate statistics because the computer can do the analysis for you only if you know which statistical test is appropriate. Students commonly come to their laboratory courses without the ability to choose the appropriate test because they have learned statistics one chapter at a time and there was only one statistical test per chapter.

We will also introduce you to the concept of a "General Linear Model". This is often neglected in an undergraduate course, but it is an important and useful concept. Based on our coverage of this model, you will be able to analyze data using the general linear model on a statistics software package (such as SPSS) should you ever need to do so. Without this understanding, you would not be able to analyze data that contains multiple variables that are collected in different ways (categorical data versus continuous data). You will be able to apply this more modern technique to appropriate datasets, and will also better understand the mathematical relationships between tests that are commonly performed in the behavioral sciences (independent t-test, paired t-test, ANOVA, linear regression), all of which you will hear about in later chapters.

Practical Details Before We Can Move On

A Short Discussion of Rounding

The rules about rounding vary a bit based on the field of study. We will follow the rule that is common in psychology, which is to round to two decimal places. For example, if you have a number such as 164.0665, we would round that number using rules that are based upon the number in the third decimal position. Keep in mind that you should not round numbers until you arrive at your final calculation. Rounding as you calculate intermediate numbers can result in the accumulation of small errors.

Rounding Rules

If the number in the third decimal position is > 5, we would round the second decimal position up by one unit. If the number in the third decimal position is < 5, we would retain the number in the second decimal. If the number was exactly 5, then we would look at the next decimal place (4th), but apply the same "less than or greater than 5" rule.

Example

For the number 164.0665, the number in the third decimal position is a 6, thus we would round the number in the second decimal position (also a 6) up one unit (7). Thus, our newly rounded number would be reported as 164.07.

However, the number 164.0652 would be reported as 164.06 since you are forced to go to the fourth decimal number and it is smaller than five. This technique may differ from the one that you were taught earlier in your education, but following this technique helps avoid bias in rounding that can occur if you round up simply based on the number in the third position being equal to five.

Buying an Appropriate Calculator

You should avoid buying a calculator that will do more than you need to do in this course. Students often buy "programmable" or "graphing" scientific calculators, and they are not only unnecessary for this course, but they often confuse the student who is trying to use them. There are a few symbols that you should look for in a calculator. The symbols sometimes vary by the manufacturer, but we will give you some of the more common variations of the symbols you need to aid you in your purchase. The calculators that we (as instructors) use for this course cost us between $6-8 U.S and were purchased in an office supply store, so you can save yourself some money as well as frustration by following our advice. You do need to find a scientific or statistical calculator and not a business calculator, but you will likely want to stay away from the programmable calculators unless you are extremely proficient with one already.

Symbols You Need On Your Calculator:

Σ or a Memory Storage Key, such as M+ or Σ+

ΣX

ΣX^2

$\overline{X}$ or μ

s_n or σ or σ_n

s_{n-1} or s or σ_{n-1}

$\sqrt{\ }$

X^2

Summary

This book will focus on the conceptual and practical aspects of statistics, particularly on how they are used in the behavioral sciences. You should purchase an inexpensive calculator that is easy to use; this will aid you in performing well in this course. In the next chapter we will introduce you to the important terminology used in statistics.

Chapter 2
Getting Started with Statistics

It's common for students to approach their introductory statistics course with some trepidation, but we believe that you will find this material easier than you expected, providing that you prepare yourself along the way. One of the best ways to do this at the beginning of the course is to thoroughly understand the basic terminology. What may appear foreign to you initially will become second nature to you given that you make yourself comfortable with the terms that are unique to this field. Our job is to make this process as easy as we possibly can.

Basic Terminology

Behavioral scientists are interested in questions such as the relationship between life stressors and domestic abuse or the number of fledglings produced in a single breeding season based on the number of helpers at the nest. In a perfect world, we would measure every instance of domestic abuse and fledgling birds, but obviously that is rarely possible. Thus, while we would like to work with an entire **population** to know the complete set of individuals, objects or scores in our research area of interest, it is much more common that we are forced to work with a subset of our population of interest. This subset is commonly termed our **sample**.

Population: The complete set of individuals, objects or scores that is under study.
Sample: A subset of the population of interest.

Figure 2.1

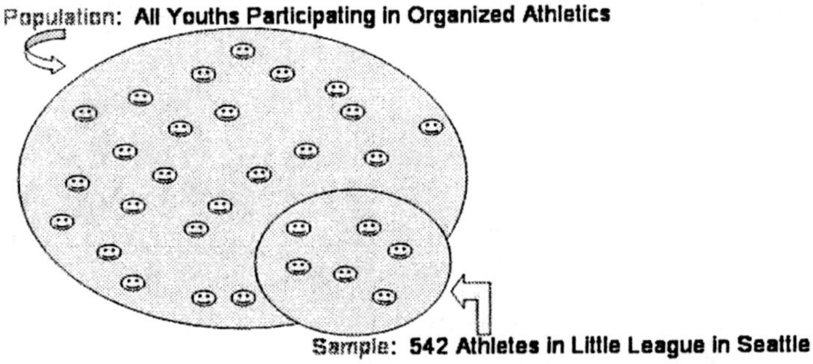

For example, Smith & Smoll (1990) studied the effect of coaching style on self-esteem in children. They were interested in the impact that various coaching styles had on children's self-esteem, but measuring every coach's behavior and the resulting impact on every child in youth sports would clearly be impossible. Thus, while their **population** is all youths that participate in sports, their **sample** was actually 542 male athletes participating in a Little League program in the Seattle area (see Fig. 2.1).

In order to study behavior, scientists must measure behavior in some quantitative or qualitative fashion. Thus, they measure properties of behavior, physiology, socioeconomic status, etc., which may take on different values at different times and change with various conditions. Some of the **variables** in the Smith & Smoll (1990) study included measures of a coach's instructiveness and supportiveness, as well as the level of self-esteem in the athlete. In contrast to a variable, a **constant** is a value which does not change under different conditions: in our discussions of statistics, **constants** will generally appear as fixed values in a given formula. **Variables** can be **discrete**, meaning that the values do not have units in between them, or **continuous**, where there are infinite values between units.

As the measurements are collected, the resulting values are commonly called **data**. These values, or **data**, can then be summarized by, for instance, taking the average or examining the high and low values, and these resulting values which summarize something about the sample are referred to as **statistics**. If you were taking the average of the scores of all of the individuals in the population of youths in sports, rather than just of the sample, you would be calculating a **parameter**, or a descriptive summary of data from an entire population.

> **Variable**: Any property that may take on different values at different times and may change with various conditions.
> **Continuous variable**: A variable that can have an infinite number of values between each unit.
> **Discrete variable**: A variable that has no values in between each unit.
> **Constant**: A value that is fixed.
> **Data:** A measurement of a variable.
> **Statistic**: A summary calculation on data collected from a sample.
> **Parameter:** A summary calculation on data collected from a population.

A classic experimental design is one is which the researcher randomly assigns some individuals to the experimental group, which receives a treatment, and randomly assigns other individuals in the sample to the control group, which experiences all aspects of the experiment EXCEPT the treatment. For example, in an effort to develop a treatment for nicotine addiction in a sample of smokers, you might assign some subjects to an experimental group that views a video on the negative psychological and physiological effects of smoking, and assign other subjects to a control group that views a video on an eye surgery. In this example, your **independent variable** is the group (control vs. treatment). Another type of independent variable might be the number of relatives of a smoker who have suffered from smoking-related illnesses. In this case, the researcher would be recording a variable which varies "naturally". Your **dependent variable**, in either case, might be the number of cigarettes smoked by the individuals in each group in the seven days following the viewing of the respective video or in the seven days following the interview to obtain "naturally varying" independent variables.

Independent Variable: A property that the researcher either manipulates experimentally or records as a naturally changing condition.
Dependent Variable: A property that the researcher measures.

Note that in the example on tobacco use, when the researcher is creating the experimental treatments (watching the videos), the researcher must **randomly assign** subjects to each group or condition. This is to ensure that subjects are equally likely to be assigned to the treatment or the control condition. This makes it less likely that the two groups will differ significantly from one another prior to watching the video. It would be a poor design if you were to allow subjects to assign themselves to each condition at their own will. Clearly, subjects that were more motivated to quit smoking prior to the study would elect to be in the treatment condition, while those who were not ready or willing to quit smoking would sign up for the control condition. This alone could mean that the two groups show a difference in the average number of cigarettes smoked in the week following the videotape, and the purpose of the experiment is to determine whether or not the treatment alone is effective in reducing smoking in a **random sample** of subjects taken from the entire smoking **population**. Not only is it necessary to **randomly assign** subjects to each condition, it is necessary to randomly select subjects from the **population** of smoking individuals. A failure to **randomly select** from the **population** as a whole could result in findings that cannot be generalized back to the entire **population** because the individuals are not typical of the **population**. It is rare for scientists to study a **sample** for its own sake: the goal of observing a **sample** while being able to generalize findings back to the population is a common one.

Descriptive versus Inferential Statistics

This book is organized so that you learn about **descriptive statistics** prior to learning about **inferential statistics**. **Descriptive statistics** are numbers that summarize your data; in other words, they are numbers that help describe your data set. These include the average (or **mean**) and the **range** (largest score minus smallest score), which you have most likely encountered in earlier mathematics, as well as the **standard deviation** and the shape of the **distribution of scores**, which you may not have encountered previously. We will save our discussion of standard deviation and the distribution of scores for the next chapter, but it is important to understand that all of these **descriptive statistics** attempt to describe a larger group of scores (your data), with a single number, such as an average, or visually, by graphing the data as a frequency distribution.

For example, the following numbers are difficult to summarize verbally, but can easily be summarized with **descriptive statistics**:

65 6 23 11 15 241 9 55 1 2 26 2 48 8 41 36 81 9

The average or central representative score, or **mean**, is equal to the sum of the scores divided by the number of scores. In this case the sum is 679 and the number of scores is 18, so $679 \div 18 = 37.72222$ (see Table 2.1). The smallest score is 1 and the largest score is 241. If you subtract the smallest score from the largest score, 241 – 1 = 240, you get the **range** of scores. The **mean** is an example of a statistic which summarizes information about where the middle of the scores is (called "Central Tendency"), while the **range** summarizes how far the scores are from the center (called "Variability"). Other statistics can summarize information about the "shape" of the scores, or the pattern of the scores around the center of the distribution of scores.

Table 2.1. Modified Excel Summary Statistics Table

Mean	37.72222
Range	240
Minimum	1
Maximum	241
Sum	679
Count	18

Inferential Statistics are numbers that we calculate based on sample data, but that we are attempting to use to generalize to a larger population of data that we have not sampled. These statistics require us not only to calculate summary statistics on sample data, but also to estimate population values that are unknown or unavailable, and then to evaluate the likelihood that the sample represents the population. The first step in learning how to perform **inferential statistics** is to learn about **descriptive statistics**, so we will save our discussion of **inferential statistics** for Chapter Five.

Descriptive Statistics: Numbers that summarize a set of data in one of three ways, including a measure of central tendency, a measure of the variability of the scores, or a measure of the shape of the distribution of scores.

Inferential Statistics: Use of sample statistics in an attempt to infer the characteristics of a population of scores.

Mean: A statistic that measures the middle, or central tendency, of a set of scores and is calculated by taking the sum of the scores and dividing by the number of scores.

Range: A statistic that measures the variability of a set of scores and is calculated by taking the largest score and subtracting the smallest score.

Statistical Notation

Do not make the mistake that many of our students have made in this course. These students have underestimated the importance of learning the statistical notation, and as a result have later felt that we were speaking to them in Greek (which, symbolically, we were!). Unfortunately, this meant that they struggled more than they should have had in order to understand the material in this book. Learning the terminology and the notation in this chapter will give you the vocabulary to understand the rest of the course.

We typically represent the variable that the scientist is measuring with an X or a Y, and the number of measurements the scientist has taken (or the size of the sample) by N or, later in the text, by the lowercase n. Subscripting is used to identify individual scores, such as X_1, X_2, X_3, etc., all the way to the last score, which can be symbolized as X_N. Now you have the tools you need to understand how to sum all of the scores. This technique is termed summation, and it simply means to add up all of the scores, or at least to add up the scores that are indicated by the notation.

The symbol for summation is $\sum$ and it indicates that you should add up the scores. If you wished to add up *all* of the scores in the sample, you would indicate that in notation in the following way:

$$\sum_{i=1}^{N} X_i$$

This notation tells you to add up all of the scores from the first score (let i = 1) to the last score (when i = N) for each individual score (X_i).
In practice, these elaborate notations are often dropped so that the notation simplifies to:

$$\sum X$$

This literally says that you should add up all of the X scores. For example,

$$\sum X = X_1 + X_2 + X_3 + \ldots X_N$$

If we use the sample data that we used to demonstrate descriptive statistics earlier in the chapter, we would calculate the sum of those scores in the following way:

$$\sum X = 65 + 6 + 23 + \ldots 9 = 679$$

Note that we have not written out every score in this equation, but you would need to add them all together, and if you do so, you will find that you get the same sum that we did when we used Excel to calculate the sum rather than calculating it manually. Refer to Table 2.2 for a list of all of the scores.

Another common use of the summation symbol in statistics is to show that the scores should be squared and then added to one another. This is most often represented as:

$$\sum X^2$$

but of course it should actually be written out in the following way to be complete:

$$\sum_{i=1}^{N} X_i^{\,2}$$

Table 2.2: Sample Data Shown Symbolically and Numerically

Symbolic Score	Actual X Score (Data)	Squared Scores
X_1	65	4225
X_2	6	36
X_3	23	529
X_4	11	121
X_5	15	225
X_6	241	58081
X_7	9	81
X_8	55	3025
X_9	1	1
X_{10}	2	4
X_{11}	26	676
X_{12}	2	4
X_{13}	48	2304
X_{14}	8	64
X_{15}	41	1681
X_{16}	36	1296
X_{17}	81	6561
X_{18}	9	81
	$\sum_{i=1}^{N} X_i = 679$	$\sum_{i=1}^{N} X_i^2 = 78{,}995$

When this command is elaborated with the scores used in the previous example, you get the following:

Symbolically

$$\sum_{i=1}^{N} X_i^2 = X_1^2 + X_2^2 + X_3^2 + \ldots X_N^2$$

With the Sample Data

$$\sum_{i=1}^{N} X_i^{\,2} = (65)^2 + (6)^2 + (23)^2 + \ldots (9)^2 = 78{,}995$$

With the Squared Values of the Sample Data

$$\sum_{i=1}^{N} X_i^{\,2} = 4{,}225 + 36 + 529 + \ldots 81 = 78{,}995$$

This is called the "Sum of the Squared X's". This concept can be confused with the **"Sum of the X's, Squared".** The second case refers to adding up all of the scores and then squaring the sum of those scores.

Symbolically

$$\left(\sum_{i=1}^{N} X_i\right)^2 = (X_1 + X_2 + X_3 + \ldots X_N)^2$$

With the Sample Data

$$\left(\sum_{i=1}^{N} X_i\right)^2 = 65 + 6 + 23 + \ldots 9 = (679)^2 = 461{,}041$$

Notice that taking the "Sum of the X's, Squared" results in a much larger number (461,041) than you get when you take the "Sum of the Squared X's" (78,995). The wording of these two sums is quite confusing, but you can dispense with that confusion by carefully noting where the squaring takes place. If the notation is simplified, as it often is in practical use, then it is a matter of differentiating between $\sum X^2$ and $\left(\sum X\right)^2$. The first symbol tells you to square the scores first, then add them together, and the second symbol tells you to add the scores together first, then square the sum! You can use the exercises at the end of the chapter to practice reading these different symbols and carrying out the appropriate operations.

Measurement Scales

Scientists often collect the type of data that can be manipulated using the notation you just learned, but they also collect data that are categorical. For example, you might be collecting data on the number of days in a month that a depressed individual felt overwhelmed or hopeless about their lives. A similar measurement of depression would be to ask individuals in your sample to rate themselves on a scale from 1-5 on how depressed they felt on average, with 1 being not depressed and 5 being very depressed. These sound like very similar measurements, and they are both attempts to measure underlying depression, but they are actually using two different measurement scales to collect the data for later statistical analyses. It turns out that there are not two, but four measurement scales that are used to collect data (dependent variable) and to design experiments (independent variable), and which ones you use determines the mathematical manipulations that you can make on the data later. This distinction is critical to understanding what inferential statistical tests you will be able to perform later. It is a concept that appears deceptively "simple" initially. Many of our students struggle more with this *simple* concept than they do with the mathematical manipulations that they expected to have difficulty in understanding.

Four Measurement Scales

Nominal: data are on a categorical, and often qualitative, scale rather than one that is quantitative.

Ordinal: data are on a categorical scale, in which categories can only be ranked in relative order.

Interval: data are collected in a manner that measures actual magnitude and which has equal intervals between possible scores, but does not have a meaningful absolute zero point.

Ratio: data are collected in a manner that measures magnitude, has equal intervals between possible scores, and contains an absolute zero point.

Examples of Each Measurement Scale

Nominal data are collected whenever, for example, you ask how many students in a class belong to each political party (Democrat, Green,

Independent, Republican, or Other) and you simply count the number (or frequency) of students that fall into each group. Your data might look something like this:

Democrat	Green	Independent	Republican	Other
31	4	8	29	1

In this example, students are placed into categories of political parties. These categories cannot be meaningfully ranked and they do not include information on magnitude (greater or less than) or a true zero point. Other examples of data collected on a nominal scale include the number (or frequency) of individuals in a sample that are wearing various types of running shoes (Adidis, Nike, Reebok, etc.) or the number of individuals in a sample or population that belong to various ethnic groups (African, Asian, Caucasian, Middle-eastern, etc.). Again, these are qualitative categories that cannot be ranked in order and do not include measures of magnitude.

Ordinal data do include information on rank or order. Some examples of this scale includes the order that runners finish in a race (1st, 2nd, 3rd, 4th, 5th, etc.). While this scale obviously tells you who performed the best in this case, it does not contain any information on the magnitude of the differences between the ranks. You do not know whether 1st and 2nd place was decided by 10 minutes or by 2/100 of a second. In the second scenario of our earlier depression example, the researchers asked subjects to rank their average level of depression on a scale from one to five. Data collected in this manner is on an **ordinal** scale and is common in the behavioral sciences, whether you are studying people or animals. An animal behaviorist might record the levels of aggression in a baboon troop by using a similar scale.

Data collected on an **interval** scale have information on order or rank like the **ordinal** scale, but the data are more quantitative because they contain information on magnitude. Thus, the scientist knows not only that one score is higher or lower than another, but exactly how much higher or lower one score is from another score. Marine biologists recording daily water temperatures (in Celsius) on the Great Barrier Reef are using an **interval** scale. Both Celsius and Fahrenheit temperature scales include information on magnitude: 23°C is warmer than 20°C, and it is warmer by 3°C. Another property of **interval** scales is that they have equal intervals between measurement points. In other words the magnitude of difference between 1°C and 2°C is the same amount of heat (or energy) as the difference between 30°C and 31°C. However, while you can measure temperatures of 0°C or 0°F, these measures are relative zero points rather than absolute zero points. What does this mean? Absolute zero is the absence of temperature, or more accurately, the absence of energy. It turns out that although 0°C is the temperature at which water freezes, it is not the absence of all heat or energy. Likewise, 0°F does not measure the absence of energy. Thus, both of these temperature scales measure on an **interval** scale, but not a **ratio** scale.

To be a **ratio** scale, they would have to contain a meaningful absolute zero point. You can measure temperature on a **ratio** scale, the Kelvin scale. In this case, 0° Kelvin is the total absence of heat or energy. Thus, it contains a meaningful zero point as well as the properties of magnitude and equal interval. Distance can also be measured on a **ratio** scale (inches, centimeters, miles, kilometers, etc.). Distance is **ratio** rather than **interval** because you can have, at least theoretically, zero distance between objects.

In the first scenario of the depression example we gave you earlier, where subjects were asked to tell the researcher how many days a month they felt depressed, we were collecting data on a **ratio** scale. In this case, reporting more days per month means that a subject feels more depressed, and these data have equal intervals (1 day) and an absolute zero point (subjects that are never depressed report zero days feeling depressed). In practical terms, this is a better measure of depression than the earlier ranked (1 to 5) scale because it contains more quantitative information that allows a statistician more choices in later statistical analyses.

Overview of Measurement Scales

We have summarized the properties of each measurement scale in Table 2.3. Some of our students have found this method of presenting the information more useful than the definitions, and many students find this table and the examples to be invaluable in understanding the practical differences between the measurements scales. Fortunately, in practice, statisticians do not discriminate between interval and ratio scales since they both contain the necessary properties for all types of inferential statistics.

Table 2.3: Summary of the Properties of the Four Measurement Scales

Scale	Order	Magnitude	Equal Intervals	Absolute Zero
Nominal	No	No	No	No
Ordinal	Yes	Some	No	No
Interval	Yes	Yes	Yes	No
Ratio	Yes	Yes	Yes	Yes

Summary

In this chapter, we have presented the basic terminology that is needed to understand the rest of the material in this textbook. We can't emphasize enough the value of learning this material very well, right now. To repeat what we said earlier, for instance, one of the seemingly easiest concepts to (apparently) learn, yet most difficult to apply, are the measurement scales. The concept of different measurement scales will become very important later in this textbook, and you will need to apply the concept to real experimental situations (the dreaded word problems!). So start learning these concepts early and learn them well. The payoff later will be significant.

References

Smith, Ronald E & Smoll, Frank L. 1990. Self-esteem and children's reactions to youth sport coaching behaviors: A field study of self-enhancement processes. *Developmental Psychology*, 26, 987-993.

Chapter Two Exercises

Explain the difference between the following terms.

1. What is the difference between a population and a sample?

2. What is the difference between a variable and a constant?

3. What is the difference between an independent variable and a dependent variable?

4. What is the difference between a statistic and a parameter?

5. What is the difference between continuous variables and discrete variables?

6. What is the difference between descriptive and inferential statistics?

Provide an answer for the following questions.

7. Why is random sampling so important?

8. Why is it important to discriminate between measurement scales?

Identify the following terms in the story problems:

- The independent variable
- The population
- The dependent variable
- The sample
- The statistic

9. A group of researchers want to test their hypothesis that sleep deprivation will decrease the number of lectures attended in a quarter. The researchers round up 350 students from the University of Washington's 38,000 undergraduates and randomly assign each student to either the 4-hours-of-sleep-per-night group or the 8-hours-of-sleep-per-night group. Each subject's attendance is recorded for 1 quarter and the mean attendance days per group is calculated.

10. 150 high school students from Santa Monica County are tested to see if teacher gender affects test scores for California high schoolers. Half of the students learn material for a test from a male teacher, and the other half learn the same material from a female teacher of equal teaching ability. Mean test scores are compared.

Define the following terms.
11. What is the nominal scale?
12. What is the ordinal scale?
13. What is the interval scale?
14. What is the ratio scale?

Answer the following questions:
15. Macy's keeps a detailed list of the different brands of shoes its sells (e.g., Doc Marten's, Nike, Nine West). What type of data is this?
   A. Nominal
   B. Ordinal
   C. Interval
   D. Ratio

16. The Kelvin scale is a(n) __________ scale while the Fahrenheit Scale is a(n) ______ scale.
   A. Nominal, Ratio
   B. Interval, Ordinal
   C. Ratio, Interval
   D. Interval, Ratio

17. Which of the following scales correctly describes its associated information?
   A. Nominal: Time it takes to finish these homework questions
   B. Ordinal: A 1-7 ranking about the helpfulness of these questions
   C. Interval: The names of different caffeine-filled drinks you might consume while trying to stay awake through these questions
   D. Ratio: Confidence level about the first test

Using the following data points, calculate the answers to the questions below:
x_1: 72 x_2: 29 x_3: 10 x_4: 49 x_5: 22 x_6: 58 x_7: 63

18a. N = ?
18b. $\sum X = ?$
18c. $\sum X^2 = ?$
18d. $\left(\sum X\right)^2 = ?$
18e. $\sum$ (X-6) = ?
18f. $\sum X + 12 = ?$

Chapter 3
The Mean and Standard Deviation

In this chapter, you will learn about some basic descriptive statistics, including the mean and standard deviation. **Descriptive statistics** are numbers that summarize your data; in other words, they are numbers that help describe your data set without giving every last raw score. **Descriptive statistics** can describe a large group of scores (your data), with a single number, such as a mean. We will begin the chapter with some descriptive statistics that indicate the average, or central tendency, of your sample, and later we will turn to measures of variability.

Descriptive Statistics: Numbers that summarize a set of data in one of three ways, including a measure of central tendency, a measure of the variability of the scores, or a measure of the shape of the distribution of scores.

Mean: A statistic that measures the middle, or central tendency, of a set of scores and is calculated by taking the sum of the scores and dividing by the number of scores.

Measures of Central Tendency

The Mean

The most common measure of central tendency, or the average, is the arithmetic mean. The **mean** is equal to the sum of the scores divided by the number of scores, and it is calculated in the same way whether you are calculating the mean of a sample or the **mean** of a population of scores.

Formula to calculate the mean of a population of scores

$$\mu = \frac{\sum X_i}{N}$$

Formula to calculate the mean of a sample

$$\overline{X} = \frac{\sum X_i}{n}$$

Example

A psychologist studying dyslexia recorded the scores from spelling tests of five children who were thought to be dyslexic based on preliminary testing. The five children had the following scores on their spelling tests which were each worth 12 points total.

Calculate the mean of this sample.

Child	Score
1	6
2	8
3	10
4	4
5	11

$$\overline{X} = \frac{\sum(6+8+10+4+11)}{5} = \frac{39}{5} = 7.8$$

$$\overline{X} = 7.8$$

Properties of the Mean

The mean has certain properties or characteristics that make it an appropriate statistic to use in summarizing a data set. These include the following:

1) The **mean** is sensitive to extreme values.

    a. Very large or very small scores will shift the **mean** so that it is less representative of the other scores in the sample or population.
    b. For example, summarizing the average salary in the Seattle Metro area by using the **mean** would be less representative if you included the multimillion dollar salary of Bill Gates. His extremely high salary would "pull" the **mean** significantly upwards and thus, it would less accurately reflect the salary of most residents of the Seattle Metro area.

2) The **mean** is sensitive to exact values.
    a. Since the **mean** is calculated by summing all of the scores and dividing by the number of scores, all values influence the calculation of the **mean**. Changing any single value (data point) by any amount will change the value of the **mean**.

3) The sum of the deviations around the **mean** is equal to zero.
    a. $\sum\left(X_i-\overline{X}\right)=0$
    b. Thus, if you subtract the **mean** from every one of the scores and then sum those "deviations from the mean", the sum will equal zero. This is logical since the **mean** should arithmetically be the middle score.

4) The sum of the squared deviations is a minimum.
    a. $\sum\left(X_i-\overline{X}\right)^2$
    b. Minimum means that the result is the smallest number that you could get. If any number other than the mean was subtracted from the scores to form the squared deviations,

the resulting sum of the squared deviations would be larger than if the mean were used. Thus, one definition of the mean is that number which produces a "minimum" sum of the squared deviations for any given data set. Take this characteristic at face value for now... we will return to it later.

5) The **mean** is the least subject to sampling variation.
   a. Thus, repeated samples result in relatively little variation in the value of the **mean**. This stability is generally a very desirable characteristic.

The Median

The **median** is the value in the middle of your scores when they are rank ordered; literally, it is the value where 50% of the scores fall below it. We can calculate the **median** of the spelling scores from above.

Median: The value at which 50% of the scores fall below it.

Example
Step One: Rank the scores from smallest to largest.

4 6 8 10 11

Step Two: Find the center score
If you have an odd number of scores, then your middle score is the **median**. In this case, it is 8. If you have an even number of scores, then you must take the average of the two center scores. For example:

2 4 6 8 10 12

In this case, both 6 and 8 are the center scores. If you take the average of 6 and 8 you get 7: $\text{Median} = \frac{6+8}{2} = 7$

Properties of the Median
Like the mean, the **median** has particular characteristics that make it an appropriate summary of central tendency in some cases. For instance:

1) The **median** is less sensitive to extreme scores than the mean.
    a. Because it is the center score, it will be less affected by very small or very large scores (like Bill Gate's annual income). Thus, you should always think about what the author is trying to summarize when the mean or the **median** is used. If someone were trying to suggest that incomes are high in Seattle, they might summarize incomes using the mean, where a few very high incomes would highly influence the mean. The **median** might be a more representative measure of central tendency in this case, because Bill Gate's income in this case is just one more value on the high side of the central tendency of the data set. The **median** is not affected by a score that's a little bit higher than the center or a lot higher. The **median** is only influenced by the number of scores on either side of the center, not their actual values.

2) The **median** is more sensitive to sampling variation than the mean.
    a. The **median** will vary from sample to sample more than the mean because the middle score will only represent one score rather than an average of all of the scores of that particular sample.

The Mode

The **mode** is the most frequent score in a sample or population, and thus it is always a real score.

Mode: The most frequent score in a set of scores.

Example
Assume that these scores are exam scores, and the total possible points are 100. Seven students took the exam.

Step One: Rank order the scores so that duplicates are easy to identify
50 72 72 88 89 93 100

Step Two: Determine any replicates in the sample
Two students earned a 72, while all of the other scores are unique. Thus, the **mode** is 72 for this sample. Note that the **mode** reflects a score that a student actually earned, whereas the **median** and **mean** may not. Since there are an odd number of scores, the **median** does reflect a true score in this case (88). However, the **mean** for this example is 80.57, which is not a true score.

Measures of Variability

Measures of variability are those that tell us how spread out the scores in a sample or population are from one another. This provides us with information that the mean does not reveal. It is possible for two samples to have the same mean, but differ in their distribution of scores (their variability) (see Fig. 3.1). It is useful to have a measure of variability so that we can determine how far a particular score is from its mean. Students often use this as a measure of how well (or how poorly) they did on an exam. Statisticians use measures of sample variation to *estimate* the variation of a population of scores that they do not have available to them.

Range

One of the easiest measures of variability to calculate is the **range**. The **range** is determined by subtracting the smallest score from the largest score, and gives one a measure of how far apart the scores of the sample or population are distributed.

Range: A statistic that measures the variability of a set of scores and is calculated by taking the largest score and subtracting the smallest score.

Example
Step One: Determine the high score and the low score.
This can be done by rank ordering the scores.

50 72 72 88 89 93 100

Step Two: Subtract the low score from the high score.
Range = High Score – Low Score
Range = 100 - 50 = 50

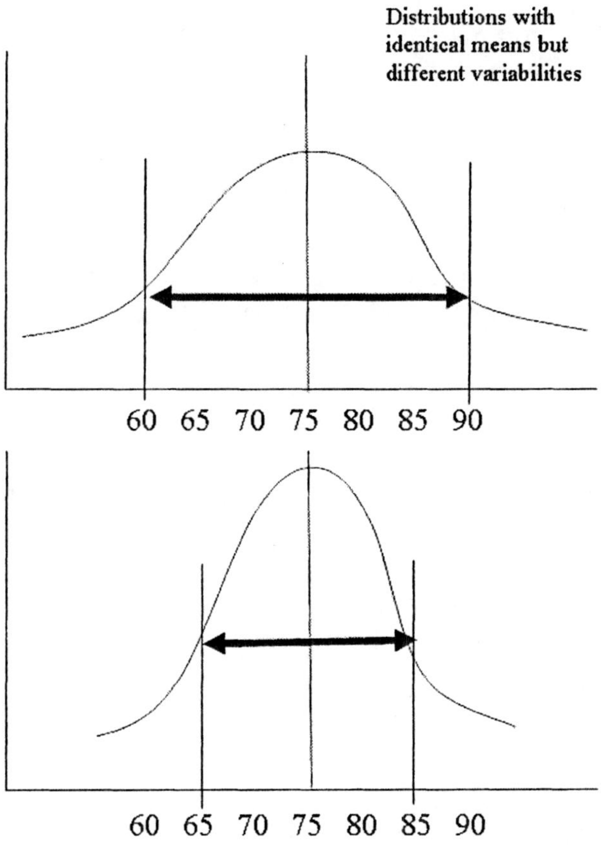

Figure 3.1

Problems with the Range as a Measure of Variability

Unfortunately, while the range is simple to calculate, it may lead to a measure of variability that is not representative of the population of scores. For example, what if you had the following scores in your sample:

1 50 50 50 50 50 50 50 50 1,000

If you calculate the range on this example, you would get 1000-1 = 999, but this is not very representative of a set of scores that are largely identical. You need a measure of variability that accounts for the distribution of *all* of the scores in the sample or population.

Standard Deviation and Variance

Both **standard deviation** and **variance** are measures of variability that measure how far away a score is from the mean of the distribution of scores: the deviation of each score from the mean. We described deviations earlier when discussing the properties of the mean. Perhaps the average of these deviations would make a useful summary of the variability of the scores about the mean. But remember that we said earlier in the chapter that the sum of the deviations around the mean is equal to zero. This clearly won't work: it will always result in a measure of variability equal to zero. But we also said that the sum of the *squared* deviations around the mean is a *minimum*, a number characteristic of that data set. Clearly, squaring the deviations before summing them provides a better measure of variability, and generally we calculate the <u>average</u> squared deviation by dividing the sum of the squared deviations by the number of scores (N).

But in trying to use this "average squared deviations" as a measure of variability, a new problem arises: the resulting squared units makes the measure difficult to interpret (what are "2.34 squared seconds" as a measure of the degree of variability of some elapsed time data?). The solution to <u>that</u> problem is to take the square root of the solution after squaring the sum of the deviations. So, the **standard deviation** is the square root of the sum of the squared deviations around the mean divided by the number of scores, or the square root of the average squared deviations. Look below to see how this idea is developed mathematically.

If you do *not* take the square root of the resulting solution and leave the answer in squared units, you have calculated the **variance**. That's right, the **variance** is left in the squared units that are more difficult to interpret descriptively, but **variance** turns out to be mathematically more useful than standard deviation for inferential statistics.

Another term for the **variance** is the "Mean Square". This refers to the fact that it is the "mean of the squared" deviations around the mean. The term "mean square" (MS) often shows up on the output from statistical software, so it is a good idea to be familiar with this alternative name and notation.

Standard deviation: the average deviation of a score from the mean.
Variance: the squared average deviation of a score from the mean.

Sum of the deviations around the mean

$$\sum\left(X_i - \overline{X}\right) = 0$$

Sum of the squared deviations around the mean

$$\sum\left(X_i - \overline{X}\right)^2$$

The sum of the squared deviations around the mean is also called the *sum of squares*. You will calculate this number often in this course as a preliminary step to calculating the standard deviation. It can be calculated the long way ($\sum\left(X_i - \overline{X}\right)^2$) or it can be calculated the fast way using the formula:

$\sum X^2 - \frac{\left(\sum X\right)^2}{n}$. We prefer the fast way since the long way is more prone to errors, but we'll show you an example worked out in both ways later on. It is important to distinguish between the sum of the squared X's and the sum of the X's squared when using the fast method. However, if you ever get a negative SS result, then you know that you have miscalculated. The most common error is to mix up the sum of the squared X's and the sum of the X's squared, which will result in a negative number.

$$\text{Sum of Squares (SS)} = \sum\left(X_i - \overline{X}\right)^2 = \sum X^2 - \frac{\left(\sum X\right)^2}{n}$$

Formula for the standard deviation of a population of scores
The standard deviation is the square root of the sum of the squared deviations around the mean divided by the number of scores. Mathematically, that is represented with the following notation:

$$\sigma = \sqrt{\frac{\sum (X_i - \mu)^2}{N}} \quad \textbf{or the fast version } \sigma = \sqrt{\frac{\sum X^2 - \frac{\left(\sum X\right)^2}{N}}{N}}$$

If you substitute the Sum of Squares Symbol for the Sum of the Squared Deviations, you can re-write the formula:

$$\sigma = \sqrt{\frac{SS}{N}}$$

The population standard deviation is represented by sigma (σ). Note also that we have used the mean of the population (μ) in this formula since we are calculating the population standard deviation. Taking the square root of the sum of the squared deviations around the mean allows us to come up with a measure of variability that is in the same units as our scores. If we left them in squared units, they would look artificially large, but if we did not square the deviations, we would always get zero, which is unlikely to represent the true variability of the scores. Thus, mathematically, we have solved our dilemma of finding a measure to accurately represent all scores and provide a minimum measure of variability. The measure is divided by the size of the population to adjust the total deviation by the number of scores in the population. This is what makes our measure an *average* deviation, which is more commonly called the **standard deviation.**

Formula for the standard deviation of a sample of scores

$$s = \sqrt{\frac{\sum \left(X_i - \overline{X}\right)^2}{n-1}} \quad \textbf{or the fast version} \quad \sqrt{\frac{\sum X^2 - \frac{\left(\sum X\right)^2}{n}}{n-1}}$$

If you substitute the Sum of Squares Symbol for the Sum of the Squared Deviations, you can re-write the formula

$$s = \sqrt{\frac{SS}{n-1}}$$

The standard deviation formula is altered slightly to apply it to samples rather than populations. The standard deviation of a sample is represented with the symbol, s, and we use the sample mean notation ($\overline{X}$) as well. The more significant change is that we are now dividing by n-1. We use the small n to indicate it is the size of a sample (rather than a population), and we subtract one from n.

Why do we use n-1 for a sample but not a population?

Mathematicians have found that when you estimate the population standard deviation from a sample standard deviation, it tends to underestimate the true variability of the population by a small amount. If you subtract one from your sample size, it provides an adjustment that ensures that your estimate of the population standard deviation is not too small. Thus, the sample standard deviation is considered a biased estimate of the population standard deviation, but the sample mean is an unbiased estimate of the population mean. If the sample mean were a biased estimate, it would require a correction.

Formula for the variance of a population of scores

The variance is the sum of the squared deviations around the mean divided by the number of scores. Mathematically, that is represented with the following notation:

$$\sigma^2 = \frac{\sum (X_i - \mu)^2}{N} \text{ or the fast way } \sigma^2 = \frac{\sum X^2 - \frac{\left(\sum X\right)^2}{N}}{N}$$

This formula is similar to the formula for the population standard deviation except that the symbol for variance demonstrates that the units are squared (σ^2) and you do not take the square root of the solution.

If you substitute the Sum of Squares Symbol for the Sum of the Squared Deviations, you can re-write the formula:

$$\sigma^2 = \frac{SS}{N}$$

Formula for the variance of a sample of scores

$$s^2 = \frac{\sum(X_i - \overline{X})^2}{n-1} \text{ or the fast way } s^2 = \frac{\sum X^2 - \frac{(\sum X)^2}{n}}{n-1}$$

This formula is similar to the formula for the sample standard deviation but without taking the square root.

If you substitute the Sum of Squares Symbol for the Sum of the Squared Deviations, you can re-write the formula:

$$s^2 = \frac{SS}{n-1}$$

Example, calculated the long way (conceptual) and the fast way
Determine the population standard deviation and variance using the conceptual formula and then the faster calculation technique.

Step One (Conceptual Formula): Determine the mean of this population of scores.

$$\overline{X} = \frac{\sum(3+4+4+5+5+5+5+6+6+7)}{10} = \frac{50}{10} = 5$$

Step Two (Conceptual Formula): Subtract each score from the mean and then square the difference (deviation):

Score	Population Mean	Difference (Deviation)	Squared Difference (Squared Deviation)
3	5	-2	4
4	5	-1	1
4	5	-1	1
5	5	0	0
5	5	0	0
5	5	0	0
5	5	0	0
6	5	1	1
6	5	1	1
7	5	2	4
Sum		**0**	**12**

Step Three (Conceptual Formula): Solve for the standard deviation using the formula.

$$\sigma = \sqrt{\frac{\sum (X_i - \mu)^2}{N}}$$

$$\sigma = \sqrt{\frac{12}{10}} = \sqrt{1.2} = 1.095$$

Step Four (Conceptual Formula): Solve for the variance using the formula.

$$\sigma^2 = \frac{\sum (X_i - \mu)^2}{N}$$

$$\sigma^2 = \frac{12}{10} = 1.2$$

Step One (Faster Formula): Calculate $\sum X^2$ and $\sum X$

Punch your data (10 scores) into your calculator using the statistical mode and retrieve $\sum X^2$ and $\sum X$ to insert into the formula or square each score individually (shown in the table below).

$$SS = \sum X^2 - \frac{\left(\sum X\right)^2}{n}$$

$$SS = 262 - \frac{(50)^2}{10} = 262 - \frac{2500}{10} = 262\text{-}250 = 12$$

Notice how we get the same value for SS (12) using this formula as we did for the conceptual formula above.

Score	Squared Score
3	9
4	16
4	16
5	25
5	25
5	25
5	25
6	36
6	36
7	49
Sum of X = 50	**Sum of X^2 = 262**

Step Two (Faster Formula): Solve for the standard deviation using the formula.

$$\sigma = \sqrt{\frac{SS}{N}}$$

$$\sigma = \sqrt{\frac{12}{10}} == \sqrt{1.2} = 1.095$$

Step Three (Faster Formula): Solve for the variance using the formula.

$$\sigma^2 = \frac{SS}{N}$$

$$\sigma^2 = \frac{12}{10} = 1.2$$

Properties of the Standard Deviation

Like the measures of central tendency, the standard deviation has certain properties or characteristics:

1) The standard deviation is a measure of variability of scores around the mean.

2) The standard deviation is sensitive to each score in the distribution. That is, since every score is included in the calculation of the standard deviation, if one score is changed at all, the value of the standard deviation is changed, unlike our first measure of variability, the range.

3) Like the mean, the standard deviation is not strongly influenced by sampling variation.

4) The standard deviation, or more specifically, variance, can be manipulated algebraically which makes it useful for inferential statistics.

Symmetry

Many statisticians prefer to get a first "feel" for their data by looking at it visually rather than with the descriptive statistics we have already discussed in this chapter. A frequency distribution graph can quickly demonstrate the range of the scores as well as the frequency of occurrence of each score in the sample or population, and can even provide a preliminary indication of the central score. If a distribution of scores is unimodal (has only one mode) and symmetrical (identical on each side of the mean), then the mean = median = mode. This is a way to tell, from the descriptive statistics, whether your data is unimodal and symmetrical, an important characteristic that will be discussed in a later chapter.

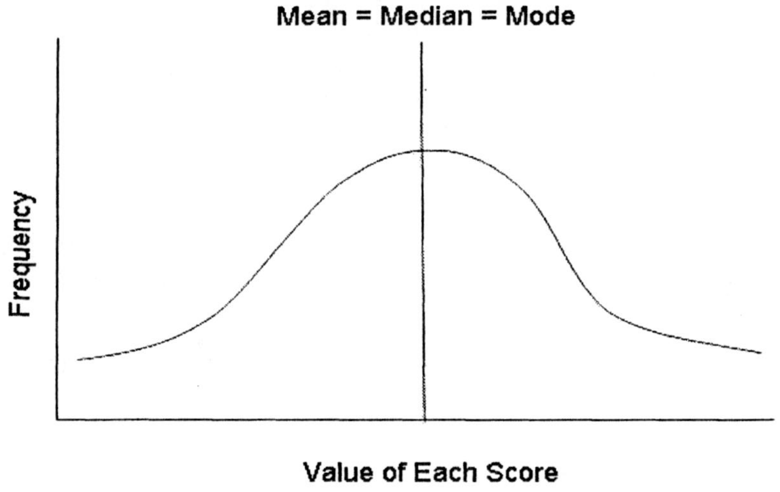

If the mean is less than the median, then the distribution has a negative skew, but if the mean is greater than the median, then the distribution has a positive skew. So skewness is one measure of the shape of a distribution of scores. There is a statistical measure of skewness, as there are for central tendency and variability, which is frequently provided with statistical software output. The measure of skewness ranges from –infinity to infinity with zero representing a symmetrical (nonskewed) distribution.

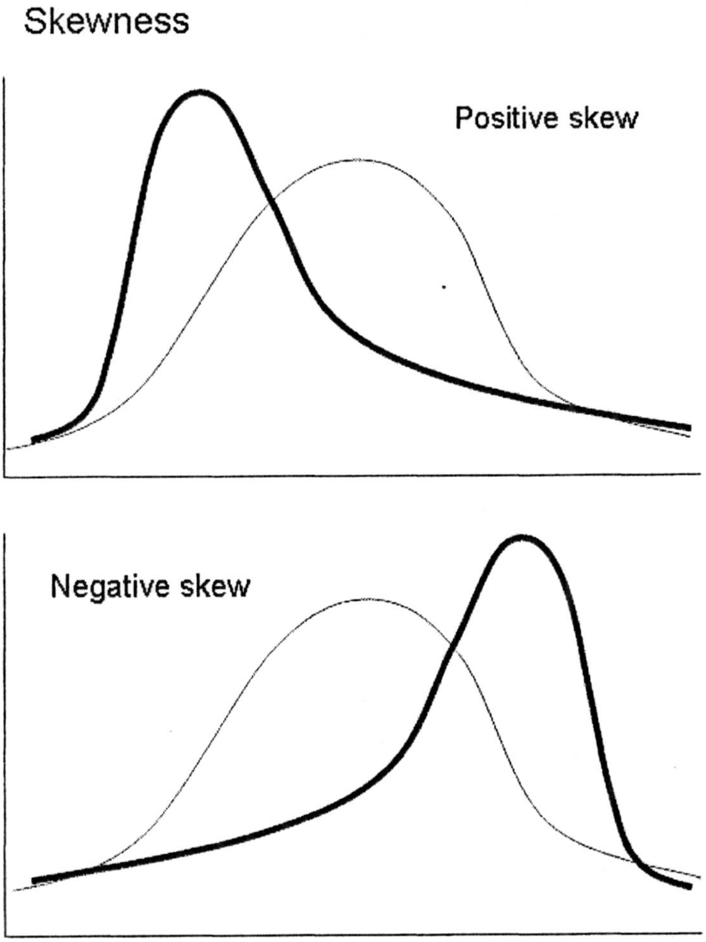

Distributions can also differ in their "peakedness". Scores may be all clustered very close to the center of the distribution (the mean) or may be very uniformly distributed about the mean. <u>Kurtosis</u> is a measure of whether or not the distribution is flat and spread out or peaked. This is more difficult to determine since the mean and median are identical in each case. Like skewness, there is a statistical summary statistic called kurtosis, which ranges from –infinity to infinity with an intermediate value of zero.

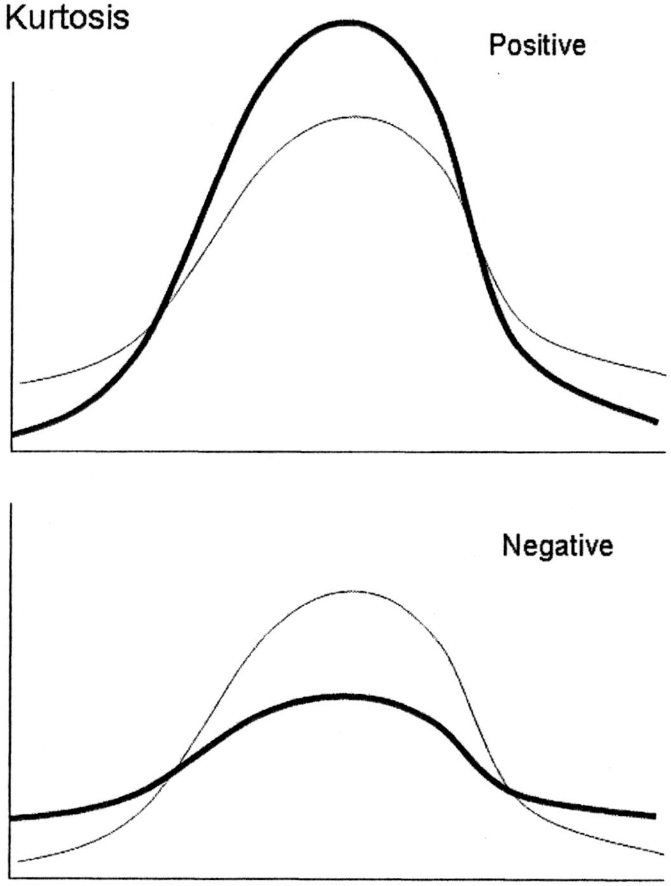

Symmetry: A distribution that is identically shaped on either side of the mean, or mirror images of one another.

Skewness: A descriptive statistical measure of whether a distribution has a higher frequency of scores on one side of the mean compared to the other side of the mean. If there are more scores on the right hand side of the mean, then the distribution is negatively skewed. If there are more scores on the left hand side of the mean, then the distribution is positively skewed. A distribution with a skewness = 0 is normally distributed and not skewed.

Kurtosis: A descriptive statistical measure of whether or not the frequency of scores in a distribution are more or less heavily clustered around the mean compared to a normal distribution. A distribution with a kurtosis = 0 is a typical bell-shaped curve ("normally distribution") and not kurtotic.

Summary

You have now completed an introduction to descriptive statistics. Many of these statistics are used to provide a summary of data, and many of them also are incorporated in determining whether or not certain assumptions of inferential tests have been met. We have enclosed a print out from Microsoft Excel demonstrating the descriptive summary statistics included in their package. Note that the measures of central tendency and many of the measures of variability you have learned are indicated in Excel.

Example: Data on the number depressive episodes shown by resident patients in a psychotic ward during one year.

Number of depressive episodes	2	1	0	7	3	11	3	0	1	1

Descriptive statistics generated by Microsoft Excel on these data.

Number of depressive episodes	
Mean	2.9
Median	1.5
Mode	1
Standard Deviation	3.510302
Sample Variance	12.32222
Kurtosis	2.514996
Skewness	1.702704
Range	11
Minimum	0
Maximum	11
Sum	29
Count	10

Chapter Three Exercises

Provide a definition for the following terms.

1. What is the mean?

2. What is the median?

3. What is the mode?

4. What is unimodal?

5. What is symmetrical?

Describe the difference between these terms.

6. What is the difference between the standard deviation and the variance?

7. What is the difference between a positive and a negative skew?

Explain the following concepts in a short answer format.

8. What do we mean when we say that the mean is unbiased but the standard deviation is biased?

9. Why is the sum of the deviations around the mean equal to zero?

10. Why do we divide the sum of the squared deviations by the sample size to calculate variance?

11. The distribution of scores of high school seniors on a <u>sixth</u> grade math test would probably be _______ .
    A. positively skewed
    B. negatively skewed
    C. normal
    D. symmetrical but not normal

12. If a large psychology class took a test on theoretical biophysics, the distribution of scores would probably be ___________.
    A. positively skewed
    B. negatively skewed
    C. normal
    D. symmetrical but not normal

13. The following sample of data represent the amount of time spent by Northwestern crows searching for food in randomly chosen 30 minute observation periods, in minutes.

    23 14 18 20 21 29 18 20

    a) Calculate the mean, sums-of-squares, and standard deviation of these data as if they are the only data that you care about, and you are not generalizing to another population.

    b) Estimate the population mean and standard deviation.

    c) Estimate the population variance.

14. A researcher has recorded the age at which a small proportion of the nursery's infant monkeys achieve the object permanence milestone of cognitive development. The data, in days, is below.

    28 25 26 29 26 25 26 27

The researcher would like to draw conclusions about the performance of monkeys raised in this lab: produce the appropriate numbers for mean and standard deviation.

15. A researcher is interested in the number of vampires killed during *Buffy the Vampire Slayer* (TV show). Out of the total population of *Buffy* shows that have aired to date, the researcher randomly selects 10 episodes and records the number of vampires killed in each episode (numbers below).

19	28	7	35	9	13
88	17	2	21	19	11

Calculate the following based on the above story:

a. Mean =
b. Median =
c. Mode =
d. N =
e. Standard deviation =

16. Professional golfers recently competed in the NEC tournament at the Sahalee Country Club in Redmond, WA. The following 9 scores are a random sample of all the scores shot during the tournament. Assume the scores are normally distributed. Find the mean, standard deviation and variance of these scores.

81	71	63	69	66	77	68	73	75

a. Mean =
b. Standard deviation =
c. Variance =

Chapter 4: The Normal Distribution, Standardized Scores & Probability

In this chapter you will learn about the normal distribution, which is a concept that you will use for the remainder of the course, and you will get your first introduction to calculating a statistic and then determining the probability of obtaining that value if only chance factors are at work. What you learn in this chapter will, for example, help explain how standard achievement tests (SATs) are interpreted so that comparisons can be made between students.

Normal Curve

The normal curve is a distribution with specific mathematical characteristics and properties. It is common for basic biological data, such as weight and height, to be "distributed normally." In fact, most popular inference tests rely on the data that you analyze being normally distributed. Thus, we would like our data to be normally distributed, or to be close enough to a normal distribution to allow us to draw inferences.

The normal curve is mathematically defined, but its formula is not necessary to your understanding of these materials. Nevertheless, it is important that you recognize its shape and properties.

Characteristics of the Normal Distribution

1) The distribution provides a description of the frequency of occurrence of each possible score (as does any frequency distribution).

2) This frequency distribution is symmetrical around the mean (each side is a mirror image of the other side) and unimodal.

3) The curve has inflection points (points at which the curve changes curvature direction) that are associated with the scores that are one standard deviation above the mean and one standard deviation below the mean.

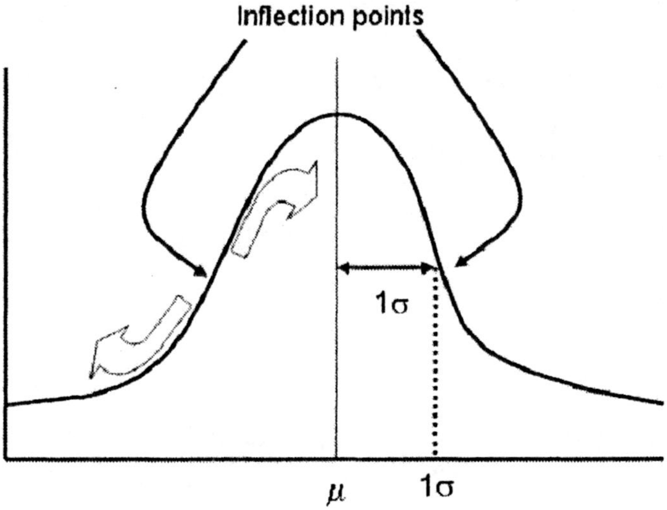

4) The total area under the curve is equal to 1.00. If you think of this as a percentage, then 50% of the area lies above the mean and 50% of the area lies below the mean, and the total area is equal to 100%. In fact, we know the area of the curve associated with each point on the horizontal line (x-axis) and can look up that area in a table (see Table A).

5) Theoretically, the tails are asymptotic, meaning that both the positive and the negative tails approach the x-axis as the positive and negative scores get more extreme, but the tails never actually reach zero frequency (the x-axis).

Standardized Scores

Standardized scores show the relative status of a score in the distribution. Knowing that someone scored a 60 on an exam means little without knowing the distribution of scores. After all, a 60 could be the low score or the high score or fall anywhere between the low and high scores. It's also important to keep in mind how far apart the scores in the distribution are from one another on average. Let's say that you scored a 60 on each of the following exams. Can you rank your performance on the three exams? It's fairly clear that Exam A was your best given the mean, but what about Exams B and C? We can place the three on the same scale and directly compare them by standardizing the scores. The result is called a z-score.

Exam A:	$\mu = 30$	$\sigma = 10$
Exam B:	$\mu = 65$	$\sigma = 20$
Exam C:	$\mu = 65$	$\sigma = 2$

Formula for a z score, or standard score calculation

$$z = \frac{X - \mu}{\sigma}$$

Notice that you subtract the population mean from the score and then divide by the population standard deviation. This formula converts scores on any scale to the same scale so that they can be directly compared. It places all scores onto a scale that is relative to the position of the center of the individual distributions (the mean) and relative to the distribution of the scores (the standard deviation). So the z score is the original score(s) converted to a score which represents the number of standard deviation units that each raw score is from its own mean. Let's work through the three exam examples to demonstrate the z score principle.

Step One: Apply the formula to each Exam Score, Mean and Standard Deviation.

Exam A: $z = \frac{60-30}{10} = \frac{30}{10} = 3.00$

Exam B: $z = \frac{60-65}{20} = \frac{-5}{20} = -0.25$

Exam C: $z = \frac{60-65}{2} = \frac{-5}{2} = -2.50$

Step Two: Evaluate the z scores based on the sign and value of the z score

Even without going to Table A, you could evaluate these three z scores. Because the z score for exam A is positive, you know that it is above the mean, or better than average. In fact it is three standard deviations above the mean, so it is far above the mean. Likewise, the z scores for exams B and C are below the mean and you know that is the case because they are negative z scores. Having a larger negative z score (as in exam C) indicates a performance far below the mean, while exam B is slightly below the mean in comparison.

Step Three: Evaluate the z scores based on the normal distribution curve

One of the most important characteristics of a normal curve is that we know the area under the curve for each value on the x-axis. We have converted the exam scores into standardized scores on a normal curve. Thus, we can look up the z values that we have calculated to determine the area under the curve for each of our three values. Because the normal distribution is symmetrical, we can look up the positive numbers and the area under the curve would be identical on the other side. Column C in Table A gives you the area under the curve from your z score into the nearest tail. Column B gives you the area between that particular z score and the mean. Column A is each z score. Begin by looking in column A until you find the z score of interest. Whether you use column B or C to find the area depends a bit on the location of the z score. It is best to draw yourself a quick picture when solving these problems so that you pick the correct column and calculate the area appropriately.

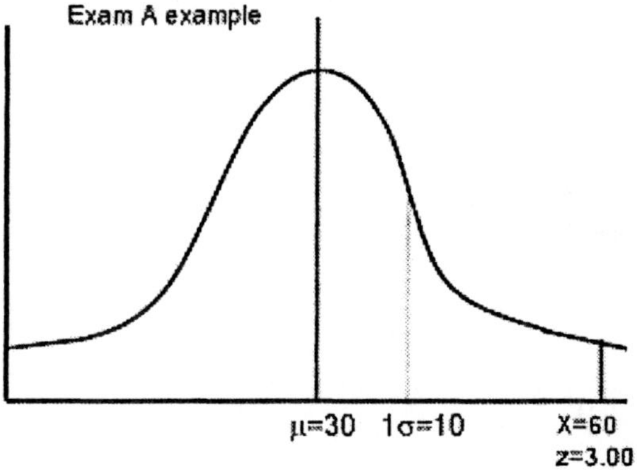
Exam A example
μ=30 1σ=10
X=60
z=3.00

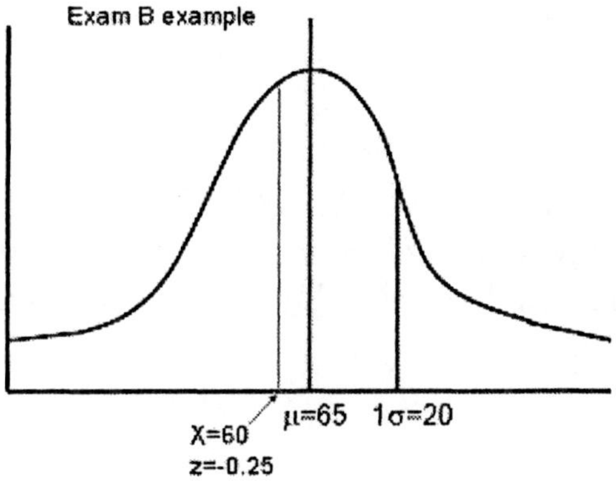
Exam B example
X=60
z=-0.25
μ=65 1σ=20

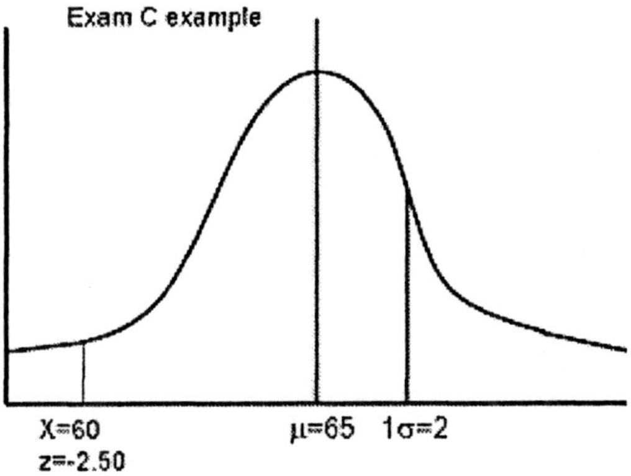
Exam C example
X=60
z=-2.50
μ=65 1σ=2

Exam A: z = 3.00 (from Table A, column C), the area beyond z is 0.0013. Thus, if you change this to a percentage (multiply by 100), you find that only 0.13% of all scores are this far above the mean or farther (more extreme than this score).

Table A, reproduced

Col A	Col B	Col C
.	.	.
.	.	.
.	.	.
2.99	.4986	.0014
3.00	.4987	.0013
3.01	.4987	.0013
.	.	.
.	.	.
.	.	.

Exam B: z = -0.25 (from Table A, column C), the area below z is 0.4013. Thus, if you change this to a percentage (multiply by 100), you find that only 40.13% of all scores are this far below the mean or further below the mean (more extreme than this score).

Table A, reproduced

Col A	Col B	Col C
.	.	.
.	.	.
.	.	.
0.24	.0948	.4052
0.25	.0987	.4013
0.26	.1026	.3974
.	.	.
.	.	.
.	.	.

Exam C: z = -2.50 (from Table A, column C), the area below z is 0.0062. Thus, if you change this to a percentage (multiply by 100), you find that only 0.62% of all scores are this far below the mean or further below the mean (more extreme than this score).

Table A, reproduced

Col A	Col B	Col C
.	.	.
.	.	.
.	.	.
2.49	.4936	.0064
2.50	.4938	.0062
2.51	.4940	.0060
.	.	.
.	.	.
.	.	.

Notice that in all three examples we used column C. This is because column C is the most intuitive to use, but we will demonstrate some problems that require the use of column B. If you refer back to Table A with a z score of 3.00, notice that column B indicates that the area between a z score of 3.00 and the mean is 0.4987 or 49.87%. If you add the area from column C to that (0.0013), you should get .5000 or 50%. This is correct because the two sides of a normal distribution are symmetrical (each contain ½ of the total area) and the total area under the curve is 1.00 or 100%. We'll go through another example to clarify this point.

Example

John scored 43 out of 60 possible points on his first introductory psychology exam. If the exam mean was 30 and the standard deviation was 8, what proportion of students scored below John and what proportion scored above John?

Step One: Apply the formula to John's exam score to calculate his z score

$$z = \frac{X - \mu}{\sigma}$$

$$z = \frac{43 - 30}{8} = \frac{13}{8} = 1.625$$

Step Two: Evaluate John's z score based on the normal distribution curve (Table A)

Look for z=1.625 in column A. You'll find a z score=1.62 and 1.63, but not 1.625. You can average the area under the curve between these two scores or just pick one to get an approximate area. If you average the area associated with a z=1.62 in column C (area = 0.0526) and the area in column C associated with z = 1.63 (area = 0.0516), you get an area beyond your z score of 0.0521. Thus, 0.0521 or 5.21% of the students scored as well or better than John on the exam. How many scored below John? If you apply the fact that the total area under the curve is 1.00 or 100%, then you can just subtract the column C area from this number to get the number of students that scored below John (1.00 - 0.0521 = 0.9479 or 94.9%). You could also use column B to solve this problem. Column B indicates that the area between John's z score of 1.625 and the mean can be averaged as 0.4479. If you add .50 or 50% to that number, then you have the total area below John. Again, we highly recommend drawing yourself a quick graph to ensure that you do not get columns B and C confused. It is as easy as confusing right and left!

Table A, reproduced

Col A	Col B	Col C
.	.	.
.	.	.
1.61	.4463	.0537
1.62	.4474	.0526
1.63	.4484	.0516
1.64	.4495	.0505
.	.	.
.	.	.
.	.	.

An Additional (more challenging) Example

The testing bureau reports that the mean for the population of Graduate Record Exam (GRE) scores is 500 with a standard deviation of 90. The scores are normally distributed. What proportion of scores lie between the scores of 460 and 600?

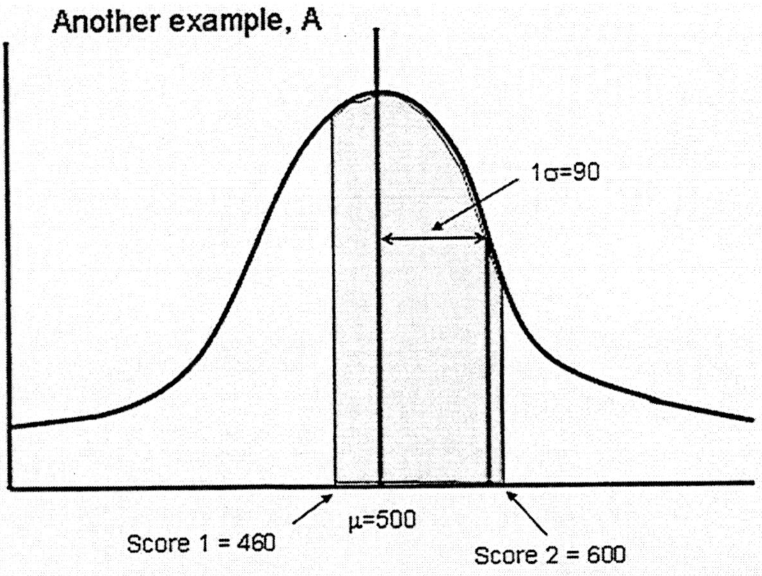
Another example, A
1σ=90
μ=500
Score 1 = 460
Score 2 = 600

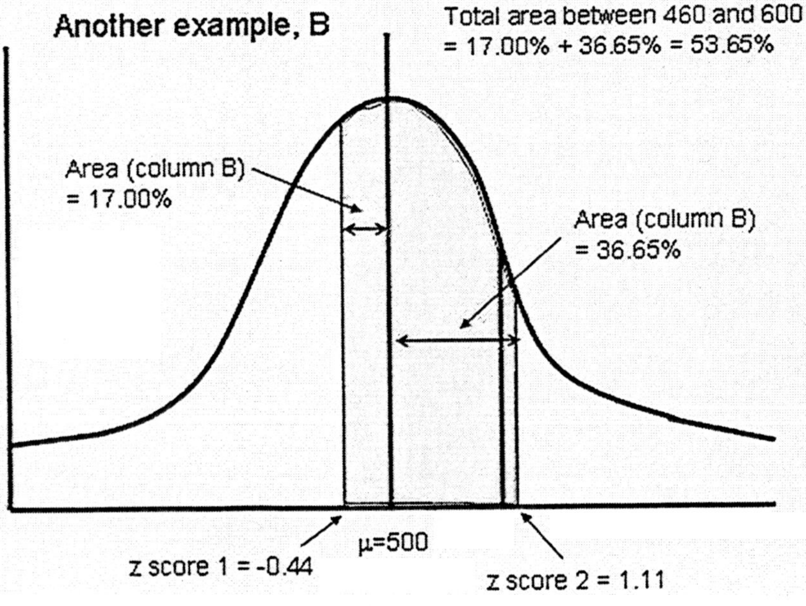
Another example, B
Total area between 460 and 600
= 17.00% + 36.65% = 53.65%
Area (column B)
= 17.00%
Area (column B)
= 36.65%
μ=500
z score 1 = -0.44
z score 2 = 1.11

Step One: Apply the z score formula to each exam score to calculate each z score

$$z = \frac{X - \mu}{\sigma}$$

$$z = \frac{460 - 500}{90} = \frac{-40}{90} = -0.4444$$

$$z = \frac{600 - 500}{90} = \frac{100}{90} = 1.1111$$

Step Two: Evaluate each z score based on the normal distribution curve
Because you want to know the area between each z score rather than beyond each z score (into the tails of the distribution), this is an ideal use of column B areas.

z = -0.4444 (from Table A, column B), the area between z and the mean is approximately 0.17. If you then change this to a percentage (multiply by 100), you find that 17% of all scores fall between this score and the mean.

z = 1.1111 (from Table A, column B), the area between z and the mean is approximately 0.3665. Thus, if you change this to a percentage (multiply by 100), you find that 36.65% of all scores fall between this score and the mean.

Step Three: Add up the areas to get the total area between the scores
(Area between z = -0.4444 and the mean) + (Area between z = 1.111 and the mean) = Total Area between z scores

0.1700 + .3665 = .5365 or 53.65% of all scores fall between GRE scores of 460 and 600.

What if you know the area, but want to find the score?

Let's say that you know the area, but want to find the score associated with that area. You might want to know what score is associated with an area of 0.75 or 75% or you might want to know what score on the Graduate Record Exam (GRE) you need to obtain in order to be in the top 10% of all applicants to graduate school. The z score formula can answer these questions as well; you just need to rearrange the formula to solve for X rather than for z.

The z score formula

$$z = \frac{X - \mu}{\sigma}$$

Rearrange the formula to solve for a score (X)

$X = \mu + \sigma (z)$

Example

If the mean of an exam is 60 and the standard deviation is 20, what score is associated with 75% of the scores being below?

Step One: Apply the rearranged formula to these parameters

$X = \mu + \sigma (z)$

$X = 60 + 20(z)$

Look in Table A, column C to find the area in column C that is associated with 0.25 (25% above or alternatively, 75% of the scores below this score).

You could also look for 0.25 in column B. You should find that z = 0.67.

$X = 60 + 20(0.67) = 73.4$

Step Two: Evaluate your calculations

A score of 73.4 is the score where 75% score below or 25% score above this value.

Probability

Probability (or its symbol, p) is a number between 0 and 1 that indicates the likelihood of an event occurring due to chance factors alone. We can combine this concept and that of normally distributed variables by using z scores. Instead of asking for the area under the curve or the percentage of individuals falling above or below the z score, we can ask for the probability of obtaining a specific z score. For example, what is the probability of obtaining a score of 98 or better on an exam where the mean is 60 and the standard deviation is 20? This should sound familiar: it's just a z score problem with the word "probability" substituted for "area" or "proportion".

Step One: Insert numbers into the z score formula and calculate z obtained

$$z = \frac{98 - 60}{20} = 1.9$$

Step Two: Evaluate the statistic

From Table A, column C, you find that the area under the curve or the likelihood of obtaining a z score of 1.9 or higher is 0.0287 or 2.87%. Using our old terminology we would say that 2.87% of the students are likely to obtain this z score or a higher z score on this exam. In our new terminology, we would say that the **probability (p)** of obtaining this score or a higher score is 0.0287, so $p = 0.0287$.

Theoretical and Empirical Probability

Probability can be written as a mathematical expression of a specific event. For example:

p (A) reads as the **probability** of the occurrence of event A or the **probability** of A.

Classical, or **theoretical probability,** is the number of events classifiable as A divided by the total number of possible events.

Theoretical probability is a priori, meaning that it is a probability that is formed on potential events, not actual events. It is the probability of rolling a die and getting a 3, which is 1/6, since there are six sides to a die. This is not the same as asking how many times a 3 came up when you actually rolled a die 300 times. This expression is an **empirical probability** rather than a **theoretical probability**.

Empirical probability is a **probability** that is calculated a posteriori (after the fact) and is the number of times A occurred divided by the total number of occurrences.

Probability: a number between 0 and 1 that indicates the likelihood of an event occurring due to chance factors alone.

Theoretical probability: the number of events classifiable as A divided by the total number of possible events.

Empirical probability: is the number of times an event classifiable as A occurred divided by the total number of occurrences.

When you have a large sample size, the **empirical probability** approaches the **theoretical probability**. Thus, the actual probabilities will be most similar to the predicted probabilities when the sample size is large, and in fact, if your sample size is infinitely large, your **empirical probability** will be equal to your **theoretical probability**. This concept will have ramifications later for the effect of the sample size on the validity of your results. Specifically, we will be more confident of the results when our sample sizes are large.

Probability Notation

As we have shown, p(A) is the notation used to suggest the **probability** of an event named A. Probabilities always range from 0 to 1.00, but it is acceptable to use fractions or decimals to express probabilities. For example, you can express **probability** as the "chances in one hundred". In scientific hypothesis testing we typically accept an error rate that is equal to "five out of one hundred", which can also be expressed as a fraction, "5/100", or a decimal "0.05."

If you have been to the horse track, you are familiar with the terminology of odds. Odds statements are ratios that are given in favor or against the likelihood of an event. For example, you might bet on a horse where the odds show 3 to 1 in favor or 3 to 1 against. You can rephrase those odds to be 3 in favor and 1 against (same as 3 to 1 in favor) and 3 against and 1 in favor (same as 3 to 1 against). Thus, in the first case, ¾ are in favor and in the second case, ¾ are against the likelihood of the event (a particular horse winning or placing in the race). In the second example, if ¾ are against, then ¼ or 25% are in favor of the event. The second case is considered "longer odds", or less likely. To easily convert odds statements into fractions, simply place the first number in the numerator (top) and then add the first and second numbers to get the denominator (bottom), and use the "in favor" or "against" wording to determine the direction of the predicted odds. For example, 8 to 1 against, is 8/9 against. Thus, the odds are $8 \div 9 = 0.888888$ that the event will not occur, or $1.00 - 0.8888880 = 0.111111$ that the event will occur.

Summary

In this chapter you have learned about the normal distribution, standardized (z) scores and probability. These concepts are the first building blocks of the infrastructure that you will need to learn inferential statistics. We will introduce more of these concepts in the next chapter. The definition of a normal distribution and the concepts of probability will be important throughout this book since many inferential tests require that your data be normally distributed so that the probability of obtaining a particular result by chance can be calculated. Thus, you will repeatedly calculate statistics on your data, but interpret them based on the likelihood of obtaining that statistic simply due to chance, not because of your experimental manipulations. If the chance (or odds) is low, then you will interpret your results as unlikely due to chance, or more likely as being the result of your experimental manipulation. We will continue this discussion in the next chapter.

Chapter Four Exercises

Give short answers to these conceptual questions.

1. What is probability in statistics?

2. What does it mean to say that the odds are 5 to 1 against?

3. What does it mean to say that the odds are 2 to 1 in favor?

4. How far would you have to go away from the mean (on a normal curve) before the curve touched the axis (i.e., before the probability of a score was 0)?

5. What is the range of values for probability?

6. How is a priori probability different than a posteriori probability?

7. If you're looking at a normal curve, at what z-scores would you find the inflection points?

8. Why can we compare sample data that have different means and standard deviations by using z-scores?

Calculate the solutions for the following problems:

9. Find z-scores for each data point in the following sample of test scores. Assume no inference is being made. 56 82 74 69 94

10a. Using a mean of 36.4 and a standard deviation of 2.7, find the score corresponding to a z-score of 3.52.

10b. With this score and the above mean (36.4), did this subject to better or worse than a subject with the same score but where the standard deviation was 6?

11. Calculate the following for a normal distribution with a mean of 54 and a standard deviation of 3.2:

    11a. A z-score for a raw score of 81:

    11b. A z-score for a raw score of 51:

    11c. What raw scores bound 60% of the distribution?

12. For an Abnormal Psychology class having an average GPA of 3.1 and a standard deviation of 0.7 ….

    12a. What percentage of students have a GPA higher than 3.5?

    12b. What percentage have between 2.5 and 3.0?

13. Let's say that the mean for the first exam in another stats class was 82 and the standard deviation was 4.8:

    13a. What is the z-score for someone scoring 91?

    13b. What did a person score if their z-score was –0.61?

14. Infant monkeys can be tested for cognitive ability in much the same way as human infants. For monkeys in a lab, the mean score for one such task is 130 with a standard deviation of 16. The most extreme 5% of the scores fall beyond which score(s)?

15. A testing bureau reports that the mean for all students who took the GRE on October 31, 2001 was 500 with a standard deviation of 90. The scores are normally distributed. What is the raw score that lies at the 90th percentile?

Chapter 5
Inferential Statistics

From this point on you will be learning about inferential statistics. Now that you understand descriptive statistics, we can move you towards making inferences about populations based on sample data.

Random sampling is a basic requirement for inferential statistics, and is necessary so that one can infer that the sample is representative of the population and thus the population characteristics can be inferred from the sample by applying the laws of probability. **Random sampling** occurs when subjects are selected in a manner that assures that each possible sample of a given size is equally likely to be selected and each subject in the population is equally likely to be selected for the sample.

Random Sampling: subjects are selected in a manner that assures that each possible sample of a given size is equally likely to be selected and each subject in the population is equally likely to be selected for the sample.

How are samples randomly selected?

Researchers often use computers or random number tables to select samples. Random number tables are effective if you pick your first number randomly, for example by closing your eyes and pointing to a number as a way to enter the table and begin.

Using a computer to select a random sample can be problematic since computers are not random, but are programmed to select numbers in a particular way, so that samples drawn with the use of a computer are pseudo-random rather than truly random. Most personal computers now use complex algorithms to manipulate an arbitrary starting value sufficiently to produce a resulting number that, for all intents and purposes, appears random.

Samples can be randomly selected in one of two ways. **Sampling with replacement** is performed such that each subject that is selected for a sample is returned to the pool prior to the next selection. **Sampling without replacement** is performed when each selected subject is not returned to the subject pool prior to the next selection. There is an interesting paradox that develops with these concepts. **Sampling with replacement** is a mathematical assumption for almost all inference tests. However, for practical purposes, most researchers use **sampling without replacement** to select subjects. In fact, many researchers think that it is better for them not to use a subject more than once in an experiment, which may be true if the measured variable might be affected by the subject awareness of experimental procedures, for instance. How do we get out of this paradox? It turns out that if the sample is very small relative to the size of the population, then the effect of **sampling without replacement** on the mathematical assumptions is minor. In fact, this assumption is routinely ignored by most researchers, but violation of this assumption (by having a large sample relative to the population size) can have serious effects on results.

Sampling with replacement: Each subject that is selected for a sample is returned to the pool prior to the next selection.

Sampling without replacement: Each selected subject does not return to the pool prior to the next selection.

Hypothesis Testing

Researchers usually conduct an experiment because they have a hypothesis that an independent variable affects a dependent variable (**alternative hypothesis**), or that there is some relationship between the variables. They wish to contrast this with the possibility that the two variables are unrelated or that any patterns that are seen are due to chance factors, rather than their experiment (**null hypothesis**). These effects can be **directional** (predicted to increase or to decrease) or **non-directional** (the direction of the effect is not stated). **Non-directional hypotheses** are more common since the researcher must have a prior scientific basis for predicting a specific direction to the effect and predicting a **directional** effect that actually turns out to be in the opposite direction means that the experiment must be repeated.

Null Hypothesis (Ho): Any effect or relationship between two variables is due to chance factors.

Alternative Hypothesis (H_A): There is an effect or relationship between two variables.

Directional Hypothesis (one-tailed): The direction of the effect is indicated in the alternative hypothesis and the null hypothesis suggests the opposite or no effect.

Non-directional Hypothesis (two-tailed): The direction of the effect is not indicated in the alternative hypothesis and the null hypothesis suggests only that there is no effect or relationship between the variables.

Researchers collect data to measure the potential relationship between two variables or the effect of one variable on another, but they can only use statistics to test whether the **null hypothesis** is false. Rejecting the **null hypothesis** because there is overwhelming information suggesting that it is false, still does not prove that the alternative hypothesis is true. If we are not able to reject the **null hypothesis**, we still cannot accept it, but we only fail-to-reject the **null hypothesis**. Thus, we can never state beyond a doubt that we have proven anything! This may make the scientific process seem pointless, but recall that part of the scientific method is in repeating experiments and testing **alternative hypotheses**. With this process, scientists are able to get closer and closer to the underlying truth. Nevertheless, because we rely on the laws of probability, and there is always a small probability that our results are due to chance factors alone, we cannot prove effects beyond all doubt. However, when specific effects are found repeatedly, and the likelihood that the effects are due to chance are small each time the experiment is conducted, the plausibility of a real effect of our independent variable becomes more and more accepted over time.

Critical Level of Probability

To test hypotheses, researchers calculate a statistic from their data which, in some way, describes the magnitude of the effect of the independent variable, and then ask how likely it would be to obtain that statistic if chance factors alone were at work. If obtaining that statistic is very unlikely under chance conditions (less than or equal to 0.05 or 5%), then the researcher can reject the null hypothesis (which states that there is no relationship between their two variables). Thus, the researcher is suggesting that if their outcome would only occur in 5% or fewer cases under normal circumstances, then the likelihood that it occurred due to chance is low and it is more likely that the variables in their experiment affected the outcome. Researchers must choose the critical level of probability for their experiment. In this case, we have chosen 0.05 or 5% because that is the most commonly accepted critical level of probability in science. The critical level of probability is called the alpha (α) level. While 0.05 is the most commonly used alpha (α) level, 0.01 (1%) and 0.10 (10%) are used fairly often as well. We will discuss the pros and cons of each level in the next section.

Types of Errors

The statistical process which we have been describing asks us to make an inference about **real effects**. Real effects are the actual effects (or perhaps lack thereof) that the independent variable has on the dependent variable, out there in the real world. Our statistics are intended to allow us to assign a probability to whether there is a real effect in our experimental design. By choosing a critical probability and comparing our calculated null hypothesis probability to that critical value, we can make a statistical decision about the existence of a real effect: there is a real effect (reject the null hypothesis) or there is not a real effect (fail-to-reject the null hypothesis). In making this decision, i.e. in calculating our null hypothesis probability, it is possible for us to make an error.

There are two types of errors that researchers and statisticians can make when using statistics to evaluate hypotheses. **Type I error** is when you reject the null hypothesis when it is actually true (you conclude that there was an effect when there was not), and **Type II error** is when you fail-to-reject the null hypothesis when it is actually false (concluding that there was no effect when there was). It sometimes helps to keep track of these mentally if you think of **Type I error** as rejecting the null when you shouldn't and **Type II error** as failing to reject the null when you should reject it. If you reject the null hypothesis when in reality it is false or you fail-to-reject the null hypothesis when in fact it is true, then you have made correct decisions (see Table 5.1).

Type I error: Error that occurs when you reject the null hypothesis when it is actually true.

Type II error: Error that occurs when you fail-to-reject the null hypothesis when it is actually false.

Real effect: when your independent variable affects your dependent variable.

Table 5.1

	Reality	
Statistical Decision	Null Hypothesis True (No Real Effect)	Null Hypothesis False (Real Effect)
Fail-to-Reject Null Hypothesis	Correct Decision	Type II error
Reject Null Hypothesis	Type I error	Correct Decision

Clearly, it is desirable to limit the likelihood of either type of error. Alpha (α) is the limit we set on **Type I error**, and beta (β) is the limit we place on **Type II error**. Recall that the critical level of probability (α) is often set at 0.05. That means that we often choose to accept a **Type I error** rate of 0.05 or less. While it might seem desirable to limit that error rate even further, the lower you make α, the higher you make the likelihood of making the opposite type of error (failing to reject the null hypothesis when it is in fact false). Thus, there is a relationship between α and β such that as α increases, β decreases. Alternatively, as α decreases, β increases. Fortunately, statisticians have found that an $\alpha = 0.05$ generally gives an acceptable β level. Let's do an example to demonstrate the concepts covered so far in this chapter.

Example

A researcher wants to know if people can solve word puzzles faster when they have ingested caffeine. He designs a study where half of a group of students first solve a word puzzle under normal circumstances. Then the same students drink several ounces of caffeinated coffee and solve another word puzzle. The other half of the group solves the puzzle under the caffeine condition first, and later solves another word puzzle when they are not under the influence of caffeine. He records the time it takes for people to solve the puzzles in both conditions. We choose to set $\alpha = 0.05$.

1. What are the researcher's null and alternative hypotheses?

2. Suppose that the researcher runs the experiment, collects and analyzes the data. He gets a p-value of 0.042, and the times were shorter with caffeine. What does this p-value mean and how should the researcher evaluate the null hypothesis?

Step One: Determine the Ho and H_A

The researcher believes that caffeine will increase the speed of solving a word puzzle. Thus, the alternative hypothesis is directional.

Ho: There is no effect of caffeine on solving a word puzzle <u>or</u> caffeine decreases the speed of puzzle solving.

H_A: Caffeine increases the speed of solving a word puzzle.

Step Two: Evaluate the Ho

<u>Decision rule</u>

If $p \leq \alpha$, then reject Ho

If $p > \alpha$, then fail-to-reject Ho

<u>Decision</u>

$p = 0.042$

$0.042 < 0.05$, so we reject Ho

Conclude that caffeine appears to increase the speed of solving a word puzzle, however there is a chance (= 0.042 or 4.2%) that we have made a type I error and falsely rejected the null hypothesis; that in fact, there is NO effect of caffeine on the speed of solving a word puzzle.

Power

Power is a measure of the ability of a statistical test to detect a **real effect** of the independent variable if the effect exists in reality. It is the probability of (correctly) rejecting the null hypothesis if it is false. The values of **power** range from 0.00 to 1.00, since it is, in actuality, a probability. Obviously it is desirable to have the highest possible **power** in your statistical work, and accordingly the highest possible probability of making a correct decision about the existence of the real effect.

There are a number of factors that influence the **power** of your statistical comparison; therefore, **power** is important from the moment you design your experiment. The issue of the **power** of your comparison will also come up if you fail-to-reject your null hypothesis. As we saw in Table 5.1, there are two possible reasons that you may fail-to-reject your null hypothesis: 1) you have made a correct decision or 2) you have made a type II error (β). The lower the **power** of your comparison, the more likely you will make a type II error (β) because:

$\text{Power} + \beta = 1.00$
Or
p(rejecting Ho if false) + p(failing to reject Ho if false) = 1.00

Values of **power** from .4-.7 may be acceptable, but .8 or higher are considered good. Thus, if you fail-to-reject your null hypothesis, but you calculate the **power** of your test and determine that the **power** was .28, you have good reason to suspect that your **power** was too low to detect a potential **real effect** of your independent variable, rather than the initial assumption that there was no real effect.

Power: a measure (between 0 – 1.00) of the ability of a statistical test to detect an effect of the independent variable if the effect exists in reality.

What effects power?

1. Magnitude of the real effect of your independent variable
   As the size of the **real effect** increases, **power** increases (↑ real effect = ↑ power).

2. Sample size
   As the size of your sample increases, **power** increases (↑ n = ↑ power).

3. Alpha (α) level
   Lowering α (which controls your chances of rejecting the null hypothesis when it is actually true) reduces your **power**, but recall that increasing α will also increase your **Type I error** rate (↓α = ↓ power and ↑β, but ↑α = ↑ Type I error).

4. Experimental Design
   Experimental designs which control for the effect of confounding variables increases **power**. For example, correctly designed correlated or repeated measures designs are more powerful than independent or between-groups designs (↓ confounding variables = ↑ power).

5. Statistical Test
   Statistical tests that require inferences about the population (parametric tests) are more powerful than statistical tests that do not require inferences (non-parametric tests). See the section below for more specifics on parametric versus non-parametric tests (parametric tests = ↑ power).

6. Type of Hypothesis
   Directional tests are more powerful than non-directional tests if you have made the correct hypothesis regarding the direction of the effect (directional hypotheses = ↑ power).

Conceptual Example

Assume that eating a candy bar raises your Intelligence Quotient (IQ) score by one point. Thus, chocolate has a small **real effect** on your IQ. IQ is measured by a standardized test where the $\mu = 100$ and $\sigma = 16$. What if you randomly sample two groups of 5 individuals from the United States? You give the first group (control group) a bagel to eat and then you give them an IQ test. You give the second group (experimental group) a candy bar to eat and then you give them an IQ test. You get the following results:

Control Group	Experimental Group
$\overline{X} = 100$	$\overline{X} = 101$
n = 5	n = 5

Notice that the mean IQ is one point higher in the experimental group compared to the control group. But, how certain are you that it is because of the candy bar and not just a random effect of whom you assigned to each group? What is the probability of getting a sample with a mean of 101 or higher when your population mean = 100? You'll learn how to calculate that in the next chapter using a modified z score formula, but we did the work for you this time and found that that p = 0.4761. That means that 47.61% of the time, you could get a sample mean of 101 or higher simply due to chance. How certain would you be that the effect was due to the candy bar if you didn't know ahead of time that candy bars increase IQ by one point? You would not be very certain, particularly since candy bars have a such a small **real effect** on IQ score. If candy bars had the effect of increasing IQ score by 50 points, then we would be more certain that the candy bar was the reason for the increased sample mean in the experimental group because the difference between the two means would be so much larger. Likewise, if you sampled 20,000 people in the control group and 20,000 people in the experimental group, we would be much more likely to believe even small differences between the two sample means could be real effects.

How do you calculate power?

To calculate power, you need to consider the factors that influence power (magnitude of the effect of the independent variable, sample size, alpha level, experimental design, statistical analysis, and whether you are testing a one-tailed or two-tailed hypothesis). In the past, most people relied on published tables and power curves to estimate power or determine an adequate sample size, but it is now more common to calculate power since a number of software packages (e.g. SamplePower 2.0 by SPSS) have been developed to make the process simple. One good website to use for power calculations is UCLA's Department of Statistics website which has a nice power calculator for multiple statistical analyses (**http://www.stat.ucla.edu/**).

Demonstration of Power

Here we provide a simple demonstration of how **power** is affected by sample size, the size of the **real effect** and alpha (α). Here we have calculated power for you, and you can observe the influence each of these factors has on a few sample experimental situations. Recall that power calculations between .4 and .7 are considered acceptable by many researchers, but that .8 or higher is considered to be a good power level for an experiment. Note that the actual power calculation values in this demonstration apply only for one statistical test.

Calculation of Power Example

Sample Size (n)	Size of the Real Effect	Alpha (α)	Power Calculation
4	All levels	0.05	0.0000
12	0.70 (moderate effect size)	0.05	0.2528
12	0.90 (large effect size)	0.05	0.8891
25	0.70 (moderate effect size)	0.05	0.5118
25	0.70 (moderate effect size)	0.01	0.3407
50	0.70 (moderate effect size)	0.05	0.8594

The table demonstrates that when your sample size is 4 and α = 0.05, there is no **power** in your experiment, regardless of the size of your **real effect**. In this situation, you would NEVER be able to reject the null hypothesis, no matter how large the real effect! However, if you increase your sample size to 12 and you have a moderate effect of your independent variable, you get a **power** result that is not zero, but is still too low. The only way to get a larger **power** with a sample size of 12 is to have a **real effect** that is larger than .70, such as .90. Unfortunately, the size of your **real effect** is not always under the researcher's control. It is much more likely that the size of the sample is under your control, thus we can increase the sample size to 25 with the same moderate **real effect** (.70) that we used with the smaller size and we find that our **power** is now approaching the acceptable range (0.5118). However, note the effect on the previous **power** calculation when you reduce your alpha level to 0.01: your **power** is diminished to 0.3407.

Parametric vs. Non-Parametric tests

Parametric statistical tests require assumptions about the parameters of the population while **non-parametric** tests do not require assumptions about the population. While all of the statistical tests you will encounter in this book require random sampling, only **parametric** tests require that the underlying population is normally distributed. This assumption can be satisfied in a number of ways, including simply knowing that the population is normally distributed, showing that the sample is normally distributed or the sample size is ≥ 30, or transforming data that are not normally distributed into normally distributed data.

Transforming raw data into normally distributed data can sometimes be accomplished by taking the logarithm of each score. Because you transform each and every score in your sample, you maintain their relative differences and can then proceed with your analyses of these log-transformed scores. Of course, the actual data points are easier to interpret, so you typically graph your raw scores, but you perform **parametric** data analyses on the transformed scores. However, log-transformation does not always result in making your data normally distributed. If transformation does not work, then you must turn to a less powerful **non-parametric** statistical test for your analyses.

When all other factors that influence **power** are held constant, **parametric** tests are more powerful than **non-parametric** tests. This is because you gain a mathematical advantage if you can assume you know something about the parameters of the population. Of course, if you assume that incorrectly, your statistical decisions may be inaccurate. In the next chapter you will learn how to calculate and evaluate your first **parametric** tests (Single sample z and Single sample t tests). You will find that these tests require knowledge of some parameters of the population (both require knowledge of μ and one requires knowledge of σ) and an underlying normal distribution.

> **Parametric tests**: statistical tests require assumptions about the parameters of the population.
> **Non-parametric tests**: statistical tests that do not require assumptions about the underlying parameters of the population.

Summary

This chapter has addressed the process that one uses to evaluate whether or not a particular experimental result is due to the experimental manipulation or due to chance. This process is dependent upon random sampling, but random sampling is not a guarantee that we will always make the correct decision about the meaning of our experimental results. Thus, there is always the chance that we have unknowingly made a Type I or Type II error, but most of the time our experiment and our analysis should be done carefully enough that we should make correct decisions. In the next chapter you will develop your first inferential statistical test.

Chapter Five Exercises

Provide a short answer for the following questions.

1. What is a "real effect"?

2. What happens to power as your alpha level changes?

3. Describe the paradox produced by the difference between sampling with replacement and sampling without replacement.

4. Why do all statistical inferences require random sampling?

5. Describe a random sample.

6. What effect does increasing your sample size have on power?

7. What are some factors that influence the power of an experiment?

8. You decide that an alpha level of .05 is too high and want to reject the null hypothesis only if the probability is 1% or less than your results due to chance. How does your decision affect the probability of making a type II error?

9. You are testing an herbal memory aid that has already been shown to have few side effects. You can only release this drug if the FDA gives you its approval. Given this situation, do you think it would be justified to set alpha at something other than .05?

10. What is the difference between a parametric and a non-parametric test?

Answer the following multiple choice questions.

11. With alpha and the effect of the independent variable held constant, as N increases, ___________
a. power increases
b. the probability of a Type I error increases
c. the probability of a Type II error increases
d. power decreases

12. With the effect of the independent variable and N held constant, as alpha becomes more stringent, ____________
a. power increases
b. power stays the same
c. power decreases
d. beta decreases

13. If H_0 is true and the probability of making a Type I error is .05, then the probability of making a correct decision is ___________
a. power
b. .05
c. .95
d. need more information

14. Maximizing the power of an experiment ___________
a. minimizes alpha
b. minimizes beta
c. increases the probability of rejecting H_0 when H_0 is true
d. increases the probability of making a Type II error

15. With other factors held constant, as the effect of the independent variable decreases, power will ____________ and the probability of a Type II error will _________
a. decrease, increase
b. decrease, decrease
c. increase, decrease
d. increase, increase

16. The power of an experiment is affected by ___________
    a. the alpha level
    b. the sample size
    c. the magnitude of the independent variable's effect
    d. all of the above
    e. a and b
    f. a and c

Answer questions 17-21 using the following information. A psychologist is interested in whether marital therapy affects the happiness of each member of the couple. Twelve couples from New York City are randomly sampled for an experiment. Each member of the couple is given an assessment where they rate their marital satisfaction and their overall happiness prior to 6 months of marital therapy. After the therapy sessions are completed, each individual fills out the assessments a second time. A statistician conducts an appropriate statistical analysis to calculate the probability that the improvement they found after 6 months of therapy could be due to chance factors rather than due to the therapy intervention.

17. What is the non-directional alternative hypothesis?

18. What is the null hypothesis that is appropriate for the non-directional alternative hypothesis?

19. If the obtained probability for this experiment is $p = 0.0384$. Using an alpha $= .05_{2\ \text{tail}}$, what should you conclude?

20. What is the population to which these results apply?

21. What error might you be making based on your conclusion in #19?

Answer questions 22-24 using the following information. Pretend you're in grad school and have just finished running a study looking at performance following negative verbal feedback. Your study used the same subjects for both conditions (before and after feedback). Now you want to look at whether negative verbal feedback had any effect on subjects' test performance.

22. What is the alternative hypothesis?

23. What is the null hypothesis?

24. If you calculate a p-value of 0.0118, did negative verbal feedback have any effect on test performance? Use an alpha level of .05.

Answer questions 25-27 using the following information. You believe students taking exams do better when the exams are taken in the afternoon than in the morning. To test your idea, you pair students in two sections of a psychology course, one in the morning and the other in the afternoon (the section instructor, GPA, and other extraneous factors are controlled for).

25. What is the alternative hypothesis?

26. What is the null hypothesis?

27. If you calculate a p-value of 0.0328, what do you conclude using an alpha level of .05?

Chapter 6
Sampling Distributions and Single Sample Tests

In this chapter you will begin calculating your first hypothesis tests. This chapter will combine the concepts you learned in Chapter Four on the normal distribution and probability with the concepts of hypothesis testing from Chapter Five. In Chapter Four you acquired the skills to compare a single score to a population that was normally distributed, and to interpret that score based on the area under the curve associated with that score or the probability of obtaining that score based purely on chance. With only a small extension of that concept, you could test a hypothesis about whether a particular score belonged to a null hypothesis population of normally distributed scores.

Single Subject z-test

Suppose the time from conception to birth in humans is approximately normally distributed, with a mean of 280.5 days and a standard deviation of 8.0 days. In a paternity case, it was demonstrated that the time from the alleged conception to the birth of a 7.0 lb (average size) baby was at least 299 days. One might hypothesize that 299 days is significantly longer than you would expect by chance, and the baby might have been conceived later than the alleged conception date. You could test this hypothesis using a single subject z-test, which is essentially what you did in Chapter Four, but we have interwoven the terminology of hypothesis testing into the same mathematical steps. Thus, you would evaluate this hypothesis in the following way using $\alpha = 0.05$:

Null Hypothesis: There is no difference between the number of gestational days for this pregnancy and the number of gestational days in all pregnant women that deliver live-born children or the number of gestational days for this pregnancy is significantly shorter than one would expect due to chance.

Alternative Hypothesis: The number of gestational days for this pregnancy is longer than the number of gestational days in all pregnant women that deliver live-born children.

Step One: Compute the proportion of women having this or a longer gestation time, which you should now recognize as the probability that this pregnancy had a gestation length that came from a normally distributed population of gestation lengths.

Formula for a z-score

$$z = \frac{x - \mu}{\sigma}$$

Applying the formula to our example

$$z = \frac{299 - 280.5}{8} = 2.3125$$

Step Two: Evaluating the probability of obtaining this score due to chance

Evaluate the z-obtained value based on alpha (α) = 0.05, and a one-tailed hypothesis.

From Table A, column C: the probability of obtaining a z-obtained score of 2.31 or more extreme from the normally distributed null hypothesis population is 0.0104, and $0.0104 < 0.05$.

Our obtained probability of 0.0104 is significantly less than our α (= 0.05), thus we **reject the null hypothesis** that suggests there is no difference is gestation length or that a shorter gestation length occurred.

How should we interpret these data in light of the paternity court case?

What would you tend to believe if you were the (statistically-knowledgeable) judge? These results suggest that about 1.04% of all normal pregnancy lengths are this long or longer. Thus, it is unlikely that the alleged conception date is accurate and it is more likely that the conception took place later than was alleged. However, there is a small probability (1.04%) that this pregnancy did indeed take place on the proposed date. This last point, that there is always a small chance of the null hypothesis being correct even when you statistically reject it, is one that you must keep in mind when interpreting the results of statistical tests.

Next we will turn to extending this concept to testing the differences between a sample and a population rather than between an individual and a population. Comparing a sample to a population is more useful to behavioral scientists than comparing an individual to a population, since we usually want to extend our results to a group rather than a single case.

Sampling Distribution of the Mean

The best way to summarize data from a sample is to look at the mean and standard deviation of a sample, and the best way to compare a sample and a population is to compare their means. Instead of comparing one score to a population mean, we want to compare a sample mean to a population mean and we want to evaluate the difference between the sample and the population means, based on chance. How do we do that? Essentially, we need to evaluate our calculated statistic (the difference between the sample and the population means divided by the standard deviation of the population) with an appropriate sampling distribution.

In other words, there are two steps to answering a question about whether or not a particular sample came from the null hypothesis population.

1) You calculate the appropriate statistic.

2) You evaluate that statistic based on its **sampling distribution.**
   If the probability $\leq \alpha$, you reject the null hypothesis, but if $p > \alpha$, then you fail-to-reject the null hypothesis.

What is a **sampling distribution**? You have worked with one **sampling distribution** already, the normal distribution, and you used it to evaluate your z-scores and then your single subject z-test.

Sampling distribution: a hypothetical representation of all possible values of a statistic with the probability of each possible value if no independent variable is acting.

Most of this textbook involves teaching you a number of statistical tests that each has its own sampling distribution, and you will learn how to calculate the statistics as well as how to evaluate them based on chance, using the corresponding sampling distribution.

How are Sampling Distributions Created?

There are three steps to creating a sampling distribution that can be used to evaluate the statistics that you calculate (see Fig. 6.1 – 6.4 for a graphical display of this concept).

1) Determine all possible samples of size N that can be drawn from the population.

2) Calculate the statistic that you are using for each sample.

3) Calculate the probability of each statistical value based on chance alone. Essentially you are determining how often each statistical value occurs by chance.

Null hypothesis population: the entire population of scores that would result if the independent variable had no effect. This (usually hypothetical) population is used to test the validity of the null hypothesis.

Sampling Distribution of the Mean: the result of drawing all possible samples from a null hypothesis population and calculating a mean for each sample, and then, based on the resulting population of means, determining the probability of obtaining each possible value of the mean.

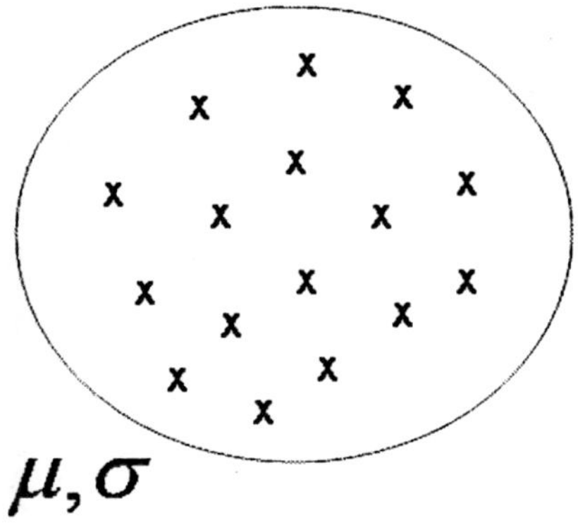

Fig. 6.1. Sampling Distribution and Standard Error of the Mean: The Underlying Raw Score Population, With Parameters Mean and Standard Deviation

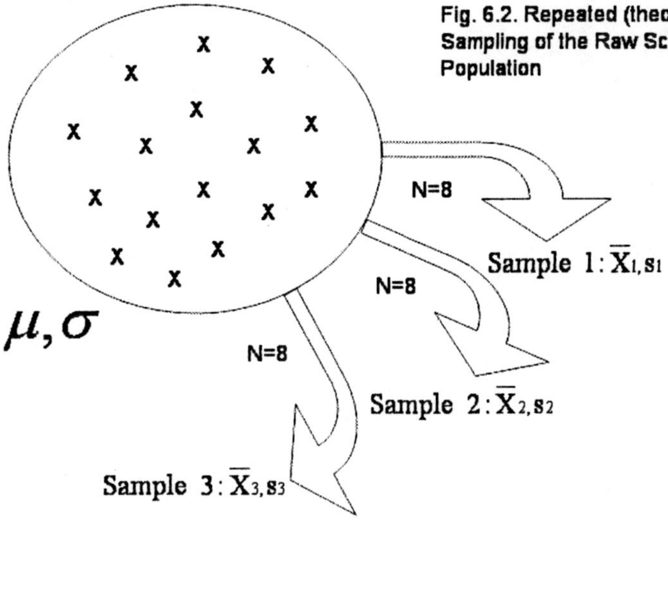

Fig. 6.2. Repeated (theoretical) Sampling of the Raw Score Population

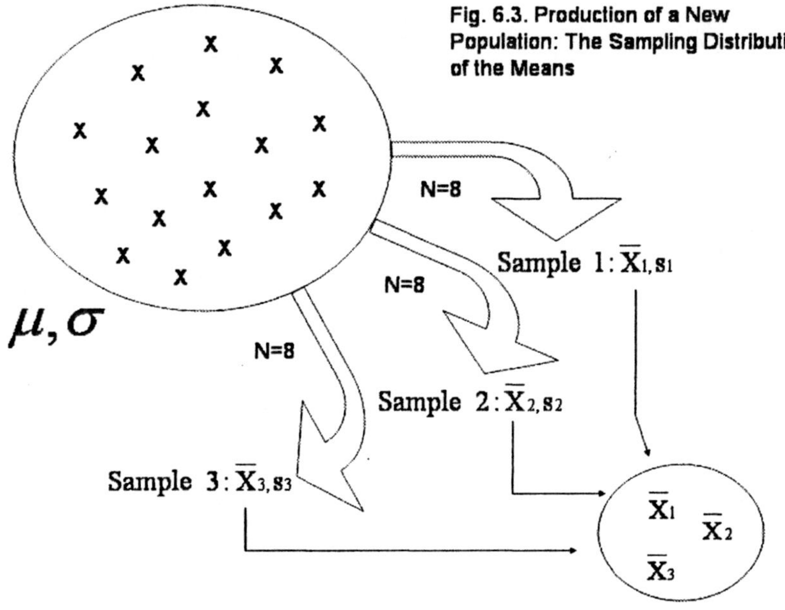

Fig. 6.3. Production of a New Population: The Sampling Distribution of the Means

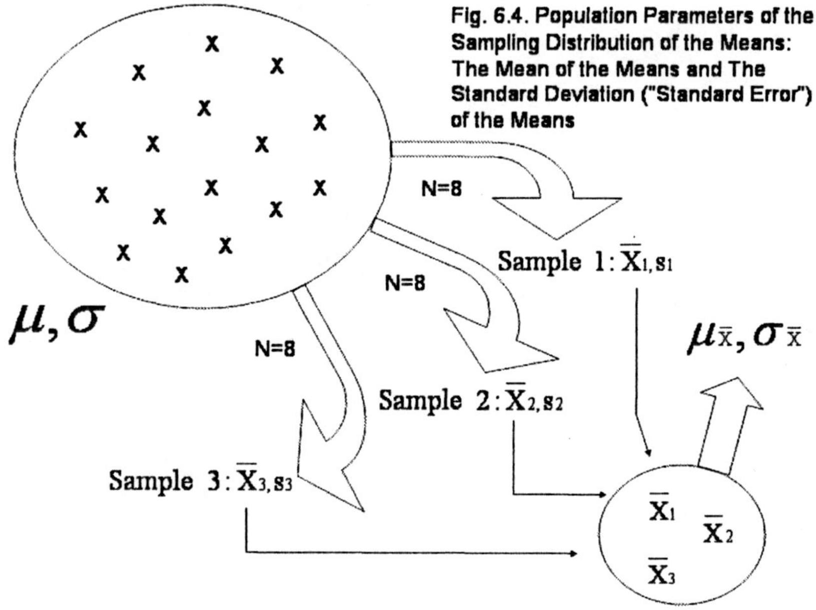

Fig. 6.4. Population Parameters of the Sampling Distribution of the Means: The Mean of the Means and The Standard Deviation ("Standard Error") of the Means

By repeating this process for different sample sizes (N) and calculating the mean of each of the samples that you randomly draw from the population, you can create a sampling distribution of the mean that has these characteristics:

1) $\mu_{\overline{X}}$ = mean of the sampling distribution of means (mean of the means).

2) $\sigma_{\overline{X}}$ = standard deviation of the sampling distribution of means. This is also called the standard error of the mean because each sample mean is an estimate of the mean of the raw score population. As an estimate, there will be some error associated with how close the estimate is from the true population values.

3) The mean of the sampling distribution of means is equal to the mean of the population of raw scores or: $\mu_{\overline{X}} = \mu$.

4) The standard deviation of the sampling distribution of means (the standard error) is equal to the standard deviation of the population of raw scores divided by the square root of N ($\sqrt{N}$), so $\sigma_{\overline{X}} = \frac{\sigma}{\sqrt{N}}$. As the sample size (N) increases, our estimate of the sample mean improves. Thus, the standard error decreases with an increasing sample size. This makes sense because if we have a population of 1,000 scores, but we draw samples of N=2, then our estimate of the mean based on each sample of N=2 is not very good and this would be reflected in our standard error. However, if we draw samples of 500, then our sample means will be close to the true population mean each time and our standard error will be smaller.

5) The sampling distribution of the mean will be normally distributed depending on the shape of the population of raw scores and on the size of our sample. Having this sampling distribution be normally distributed is a very handy feature, since we have seen how easy it is to work with probability and normal distributions. First, the sampling distribution of the mean will be normally distributed, regardless of the

> sample size used to create it, if the underlying raw scores are normally distributed. But if the raw scores are not normally distributed, or if we don't know that they are, there is still hope. The Central Limit Theorem (a mathematical proof of which you don't need the details: trust us!) tells us that regardless of the distribution of our raw scores, the sampling distribution of means will approach a normal distribution as our sample size increases. Typically, we assume that it will be normally distributed if our sample size is greater than or equal to 30 ($N \geq 30$). This assumption can be satisfied in a number of ways, including knowing that the population is normally distributed, showing that the sample is normally distributed or the sample size is $\geq$ 30, or transforming data that are not normally distributed into normally distributed data by taking the Log of each score (as we discussed in the previous chapter).

Do you see how a normal distribution could have been created by calculating all possible values that z-obtained could take and then by assigning the likelihood of each value by how often it occurs? If not, review Figures 6.1 – 6.4 as a visual aid and imagine that we are taking each raw score and subtracting the population mean from it and then dividing by the population standard deviation. Imagine repeating that for every possible score. That would create a wide range of z-obtained values, some of which would be common (those would come from scores close to the mean) and others would be less likely (those would be far from the mean). We could assign probabilities based on how often each z-score came about by random sampling of the raw scores. Figures 6.1 – 6.4 are demonstrating this same concept except that we are randomly drawing samples and then calculating the mean of the sample rather than calculating z-scores. Nevertheless, the process is to randomly sample from the population, calculate a statistic and determine the probability of that statistic occurring (whether it is a z-score or a mean). We are now ready to calculate the normal deviate z test, which is also called the single sample z test.

Single Sample z test

In this section we will convert the single subject z test to a single sample z test, which requires us to substitute some of the terms. Note the minor changes that we make to the z score formula.

Raw Score Formula

$$z = \frac{x - \mu}{\sigma}$$

Mean Score Formula

$$z = \frac{\overline{X} - \mu_{\overline{X}}}{\sigma_{\overline{X}}}$$

Mean Score Formula Simplified

$$z = \frac{\overline{X} - \mu}{\sigma_{\overline{X}}}$$

Let's do a simple example with the numbers only – we'll do another example with a word problem later on.

Step One: Evaluate the values you are given (and whether or not you need to make additional calculations prior to plugging the values into the z formula).

Let's assume that you have drawn a sample from a population, and the sample mean ($\overline{X}$) is equal to 72. You know that the population mean (μ) is equal to 75 and the standard deviation of the sampling distribution of means (or standard error) is equal to 1.66. Remember that the formula for the standard deviation of the sampling distribution of means ($\sigma_{\overline{X}}$) is equal to $\frac{\sigma}{\sqrt{N}}$, but we have given you $\sigma_{\overline{X}}$ in this example, so it is not necessary to divide by the square root of N this time.

Applying the formula to our example

$$z = \frac{72 - 75}{1.66} = -1.88$$

Step Two: Evaluating the probability of obtaining this score due to chance

Evaluate the z obtained value based on alpha (α) = 0.05, and a one-tailed hypothesis.

From Table A, column C: the probability of obtaining a z value of -1.88 (just look up the absolute value of 1.88 in the table) from the normally distributed null hypothesis population is 0.0301, and 0.0301 < 0.05.

Our obtained probability of 0.0301 is less than our α of 0.05, thus we **reject the null hypothesis** that suggests there is no difference between our sample and the population or that the difference is in the opposite direction that was outlined in the alternative hypothesis. We conclude that it is unlikely that a sample with a mean of 72 comes from our null hypothesis population that has a mean of 75.

More Complete Example

The Washington National Primate Research Center has maintained birth weight records for 40 years on 600 newborn pigtailed macaque monkeys. The birth weights are normally distributed, with a mean of 425 grams and a standard deviation of 60 grams. In the 1990's they were trying to determine the effects of AZT (*Azidothymidine*) on fetal development. AZT is a drug used to treat HIV infection (Human Immunodeficiency Virus), but at that time it was unknown whether HIV-positive pregnant mothers could take the drug without harming their unborn children. Four pregnant pigtailed macaques were given AZT during pregnancy. None of the monkeys were infected with HIV, so the study was to determine the effects of AZT alone on fetal development. One of the measures used to assess the effect of AZT on fetal development was birth weight, and those scores are listed below.

Subject	Birth weight (grams)
1	340
2	410
3	365
4	315

Null Hypothesis: There is no difference between the mean birth weight of a sample of AZT- exposed pigtailed macaques and mean of the null hypothesis (untreated) population.

Alternative Hypothesis: There is a difference between the mean birth weight of AZT-exposed pigtailed macaques and the null hypothesis population.

Step One: Compute the probability of the mean birth weight of this sample given that the sample comes from the null hypothesis population of normally distributed birth weights.

Formula for a z test

$$z = \frac{\overline{X} - \mu}{\sigma_{\overline{X}}}$$

Substitution of $\sigma_{\overline{X}}$

$$\sigma_{\overline{X}} = \frac{\sigma}{\sqrt{N}}$$

Calculate the mean of the sample

$$\overline{X} = \frac{(340 + 410 + 365 + 315)}{4} = 357.5 \text{ grams}$$

Applying the formula to our example

$$z = \frac{357.5 - 425}{\frac{60}{\sqrt{4}}} = \frac{-67.5}{30} = -2.25$$

Step Two: Evaluating the probability of obtaining this score due to chance

Evaluate the z-obtained value based on alpha (α) = 0.05, and a two-tailed hypothesis.

From Table A, column C: the probability of obtaining a z-obtained score of -2.25 (look up the absolute value of 2.25 in the table) from the normally distributed null hypothesis population is 0.0122. However, the hypothesis we are evaluating in this case is two tailed and the table only gives you the area under the curve (or probability) for one tail, thus we need to double that probability.

Double the probability for two-tailed evaluation

0.0122 X 2 = 0.0244

Compare the two-tailed probability with our alpha level

$0.0244 < 0.05$

Our obtained probability of 0.0244 is less than our α of 0.05, thus we **reject the null hypothesis** that suggests there is no difference in birth weight.

Alternatively, you could compare a **z-critical value** with your z-obtained value. When α = 0.05 and your hypothesis is two tailed, you should use the z-critical value that lists 0.025 in column C (doubled, that would be 0.05). Thus, the z-critical values for this example would be +/- 1.96.

To reject the null hypothesis, the absolute value of z-obtained must be equal to or more extreme than (greater than) the z-critical.

$|Z_{obtained}| \geq |Z_{critical}|$

$|-2.25| \geq |1.96|$, so we reject the null hypothesis.

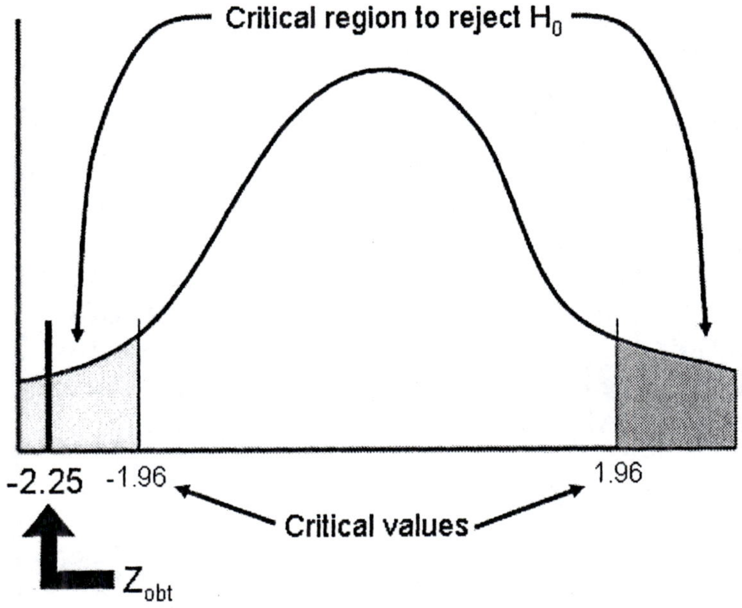

Figure 6.5

How should we interpret these data in light of the risks of AZT on fetal development?

Based on a sample size of 4, these results suggest that only about 2.44% of all samples of normal pregnancy birth weights would have a similar mean birth weight. Therefore, it is unlikely that this sample of birth weights come from the normal null hypothesis population of birth weights, and it is more likely that AZT affects birth weight. However, there is a small probability (2.44%) that these are normal birth weights that were not affected by AZT.

How would these results be reported in a scientific journal article?

If the journal required American Psychological Association (APA) format, the results would be reported in a format something like this:

> There was a significant difference between the mean birth weight of AZT-exposed pigtailed macaques and the null hypothesis population, $z(N=4) = -2.25$, $p = .02$.

This formal sentence includes the dependent variable (birth weight), the independent variable (AZT-exposed or non-AZT-exposed), as well as a statement about statistical significance, the symbol of the test (z), the sample size, the statistical value (-2.25), and the exact probability of obtaining this result simply due to chance (.02).

Keep in mind that the specific journal (Child Development, Journal of Personality and Social Psychology) may not strictly adhere to the APA standards due to editorial convenience. Journals such as Animal Behaviour or The British Journal of Social Work may deviate more dramatically from APA format.

A Second Complete Example

An industrial psychologist has been hired to advise on the establishment of a new manufacturing facility, a facility which must exhibit maximum efficiency of its workers to be competitive in the marketplace. To establish some baseline information, the psychologist measures job efficiency for four manufacturing lines, a measure for which there is considerable established data from other manufacturing facilities. In fact, it has been established that, industry-wide, efficiency has a mean of 2.6 widgets per day per line, with a standard deviation of 0.3 widgets per day per line. He collects the following data on widget production in the new plant:

Manufacturing Line	Widget Production, per day
1	2.8
2	3.2
3	3.4
4	2.6

Null Hypothesis: There is no difference between the mean efficiency of this factory's manufacturing lines and the mean efficiency of the null hypothesis population (established manufacturing lines in this industry).

Alternative Hypothesis: There is a difference between the mean efficiency of this factory's manufacturing lines and the mean efficiency of the null hypothesis population.

Step One: Compute the probability of the mean efficiency of this sample given that the sample comes from the null hypothesis population of normally distributed efficiencies.

Formula for a z test

$$z = \frac{\overline{X} - \mu}{\sigma_{\overline{X}}}$$

Substitution of $\sigma_{\overline{X}}$

$$\sigma_{\overline{X}} = \frac{\sigma}{\sqrt{N}}$$

Calculate the mean of the sample

$$\overline{X} = \frac{(2.8 + 3.2 + 3.4 + 2.6)}{4} = 3.0 \text{ widgets per day per line}$$

Applying the formula to our example

$$z = \frac{3.00 - 2.60}{\frac{0.30}{\sqrt{4}}} = \frac{0.40}{0.15} = 2.67$$

Step Two: Evaluating the probability of obtaining this score due to chance

Evaluate the z-obtained value based on alpha (α) = 0.05, and a two-tailed hypothesis. From Table A, column c: the probability of obtaining a z-obtained score of 2.67 (look up 2.67 in the table) from the normally distributed null hypothesis population is 0.0038. However, the hypothesis we are evaluating in this case is two tailed and the table only gives you the area under the curve (or probability) for one tail, thus we need to double that probability.

Double the probability for two-tailed evaluation

0.0038 X 2 = 0.0076

Compare the two-tailed probability with our alpha level

$0.0076 < 0.05$

Our obtained probability of 0.0076 is less than our α of 0.05, thus we **reject the null hypothesis** that suggests there is no difference in birth weight.

Alternatively, you could compare a z-critical value with your z-obtained value. When α = 0.05 and your hypothesis is two tailed, you should use the z-critical value that lists 0.025 in column c (doubled, that would be 0.05). Thus, the z-critical values for this example would be +/- 1.96.

To reject the null hypothesis, the absolute value of z-obtained must be equal to or more extreme than (greater than) the z-critical.

$|Z_{obtained}| \geq |Z_{critical}|$

$|2.67| \geq |1.96|$, so we reject the null hypothesis.

How should we interpret these data in light of the efficiency of these new manufacturing lines relative to industry standards?

Based on a sample size of 4, these results suggest that only about 0.76% of all samples of industry-standard manufacturing lines would have a similar mean efficiency. Thus, it is unlikely that this sample of line efficiencies come from the normal null hypothesis population of manufacturing efficiencies, and it is more likely that this factory's manufacturing line efficiencies depart significantly from industry standards. However, there is a small probability (0.76%) that these are normal efficiencies and that the differences that we observed (mean efficiency of 3.00 rather than 2.60) are simply due to chance.

How would these results be reported in a scientific journal article?

If the journal required American Psychological Association (APA) format, the results would be reported in a format something like this:

> There was a significant difference between the mean efficiency of this factory's manufacturing lines and the mean efficiency of the null hypothesis population, $z(N=4) = 2.67, p = .01$.

This formal sentence includes the dependent variable (mean efficiency), the independent variable (this factory versus population of factories), as well as a statement about statistical significance, the symbol of the test (z), the sample size, the statistical value (2.67), and the approximate probability of obtaining this result simply due to chance (.01). The approximate probability is given in this case because APA format requires that you round to two digits.

When is it appropriate to use the z test?

1) When the experiment involves a single sample mean and the parameters of the corresponding null hypothesis population are known (μ and σ), <u>and</u>

2) When the sampling distribution is normally distributed, which is the case if:
   a. The sample size is greater or equal to 30 ($N \geq 30$) or,
   b. The null hypothesis population of raw scores is known to be normally distributed.

Unfortunately, it is not very often that scientists know the mean and the standard deviation of the population. It is a little more common to know the mean of the population, but not the standard deviation. One example of this situation is when you can assume that the mean of the population is zero, but the standard deviation of the population is unknown. For example, you might assume that the average genetic relatedness of students in a random undergraduate statistics course is 0.00, but you could actually measure the relatedness through DNA sampling and determine whether or not the sample comes from the null hypothesis population of unrelated individuals. In this case, we "know" the mean of the null hypothesis population (0.00) but not the standard deviation, and we know the mean and standard deviation of the sample (of relatedness values determined by DNA sampling). We would like to compare the relatedness determined in our sample (the sample mean) to the population mean when we assume no relatedness (the population mean = 0.00), but we only know the standard deviation of the sample, not the population. In this case, a modification of the Single Sample z test is appropriate, and it becomes the Single Sample t test.

Single Sample t test

In this situation we want to compare a sample to a population, but we do not know the standard deviation of the population. The solution to this problem is to estimate the standard deviation of the population from the sample standard deviation. Remember that we used substitution to alter the single subject z score to a single sample z test. We will do that again to convert the single sample z test to a single sample t test, but this time we will substitute the sample standard deviation (s) for the population standard deviation (σ).

Comparison of Single Sample z and Single Sample t Formulas

$$z = \frac{\overline{X} - \mu}{\sigma_{\overline{X}}} = \frac{\overline{X} - \mu}{\frac{\sigma}{\sqrt{N}}}$$

$$t = \frac{\overline{X} - \mu}{s_{\overline{X}}} = \frac{\overline{X} - \mu}{\frac{s}{\sqrt{N}}}$$

So, just as $\sigma_{\overline{X}} = \frac{\sigma}{\sqrt{N}}$, it is true that $s_{\overline{X}} = \frac{s}{\sqrt{N}}$. It may be necessary in some problems to calculate the sample standard deviation (s) from raw scores. You will find the following formulas, from Chapter 3, helpful should you need to do that.

$$s = \sqrt{\frac{SS}{n-1}} \qquad SS = \sum X^2 - \frac{(\sum X)^2}{n}$$

The t critical values are dependent on sample size (n)

Because we are now going to be estimating σ from s, and the accuracy of that estimation is dependent on n (Chapter 3), our critical probability values are now going to vary with our sample size, unlike the z distribution in which the probabilities were independent of sample size. Specifically, we will use a new distribution called the t distribution. It will vary with **degrees of freedom** (df), which are related to sample size.

Degrees of freedom: the number of scores that are "free to vary" when calculating a statistic. The remaining value or values are then fixed.

The degrees of freedom for the standard deviation are equal to n-1.

The reason for this is that there is an outside constraint. You learned in an earlier chapter that the sum of the deviations around the mean is always equal to zero. If that is the case and you know all of the scores except the last one, then the last score is fixed. It must be the score that is necessary to make the sum of the deviations around the mean equal to zero. Thus, there is a constraint on the last value. For example,

Score (X)	Sample Mean ($\overline{X}$)	Deviation
3	6	-3
5	6	-1
10	6	***FIXED VALUE = 4***
$\sum X = 18$		$\sum$ Deviations = 0

Note that the last deviation must be equal to 4 in order for the sum of the deviations to equal zero. Thus, the last score is not free to vary, but must take the value that will lead to the sum of the deviations being equal to zero. This may seem a bit mysterious to you, but there is a reason for bringing up this issue. When we calculate t-obtained, we must first calculate the standard deviation. Thus, we lose one degree of freedom each time we calculate s. Therefore, the single sample t test has n -1 degrees of freedom and this is what determines your t-critical value. Let's move on to an example to clarify.

Complete Single Sample t test Example

Let's repeat our example with AZT and birth weight in pigtailed macaques, but this time, let's assume that we do not know the standard deviation of the population. The population of birth weights is normally distributed, with a mean of 425 grams. A sample of four newborn macaques that were exposed to AZT had the following birth weights:

Subject	Birth weight (grams)
1	340
2	410
3	365
4	315

Null Hypothesis: There is no difference in the birth weights of a sample of AZT- exposed pigtailed macaques and the null hypothesis population.

Alternative Hypothesis: There is a difference in the birth weights of AZT-exposed pigtailed macaques and the null hypothesis population.

Step One: Compute the probability of the mean birth weight of this sample given that the sample comes from the null hypothesis population of normally distributed birth weights.

Formula for a t test

$$t = \frac{\bar{X} - \mu}{s_{\bar{X}}}$$

Substitution of $s_{\bar{X}}$

$$s_{\bar{X}} = \frac{s}{\sqrt{N}}$$

Calculate the mean of the sample

$$\bar{X} = \frac{(340 + 410 + 365 + 315)}{4} = 357.5 \text{ grams}$$

Calculate the Sums of Squares

$(\sum X)^2 = (340 + 410 + 365 + 315)^2 = 2{,}044{,}900$

$\sum X^2 = (340)^2 + (410)^2 + (365)^2 + (315)^2 = 516{,}150$

Calculate the standard deviation of the sample

$$s = \sqrt{\frac{SS}{n-1}} \qquad SS = \sum X^2 - \frac{(\sum X)^2}{n}$$

$$SS = 516{,}150 - \frac{(1430)^2}{4} = 516{,}150 - 511{,}225 = 4{,}925$$

$$s = \sqrt{\frac{4{,}925}{4-1}} = \sqrt{\frac{4{,}925}{3}} = \sqrt{1641.666667} = 40.51748594$$

Applying the formula to our example

$$t = \frac{357.5 - 425}{\frac{40.5}{\sqrt{4}}} = \frac{-67.5}{20.25} = -3.33$$

Step Two: Evaluating the probability of obtaining this score due to chance

Evaluate the t-obtained value based on alpha (α) = 0.05, and a two-tailed hypothesis. To evaluate your t-obtained value, you must use a new sampling distribution, which is the t distribution. You will find the t distribution table in the back of your book as Table B. To determine your t-critical value, you will need to know your alpha level (0.05), the number of tails you are evaluating (2 in this case) and your degrees of freedom (N-1 = 4-1 = 3). Because the critical value varies with degrees of freedom, a complete t distribution table would take up an entire textbook. Due to limited space, you will find only a limited number of t critical values which are the most commonly used. You will not be able to determine the exact probability of obtaining each t-obtained value that you calculate. You will only know the exact probability of a given t-obtained value if it happens to be the same as a t-critical value that is listed in Table B.

Thus, when performing hand calculations that you evaluate based on statistical tables, you are limited to determining only whether the probability of obtaining your statistical result is less than or greater than your alpha level. Fortunately, numerous software packages are available to calculate statistical tests, and these software packages contain the entire set of tables. You will be exposed to two of these software packages in future chapters so that you can interpret their probabilities simultaneously with examples worked out via manual calculations.

Using a t critical value to evaluate your results

Compare a t-critical value with your t-obtained value. When $\alpha = 0.05$, your degrees of freedom are equal to 3, and your hypothesis is two tailed, you should use 3.182 as your t-critical value.

To reject the null hypothesis for a two-tailed test, the absolute value of t-obtained must be equal to or more extreme than your t-critical.

$$|t_{obtained}| \geq |t_{critical}|$$

$|-3.33| \geq |3.182|$, so we reject the null hypothesis.

How should we interpret these data in light of the risks of AZT on fetal development?

These results suggest that something less than 5% of all normal pregnancies have similar birth weights. Thus, it is unlikely that these birth weights come from the normal null hypothesis population of birth weights and it is more likely that AZT affects birth weight. However, there is a chance (<5%) that these are normal birth weights that were not affected by AZT.

When is it appropriate to use a Single Sample t test?

1) Use a Single Sample t test when you have one sample; you know the population mean but not the population standard deviation; you have an interest in comparing your sample mean to the population mean, and

2) When the sampling distribution is normally distributed, this is the case as earlier when:
   a. The sample size is greater or equal to 30 ($N \geq 30$) or,
   b. The null hypothesis population is known to be normally distributed.

How to Choose the Appropriate Test

Knowing how to choose the appropriate statistical test is arguably the most important skill you will learn in this course. Computers are better at calculating the tests than they are at determining the correct test for the design and measurement scale. Thus, you will need to hone your skills in choosing the appropriate test as you proceed through this course. To aid you in this process, we have provided a decision flowchart that will become more complicated as you learn more statistical tests. Right now it is fairly simple, but becoming familiar with it early will be a great aid to you, and you can use it to solve the "choose the appropriate test" questions in the exercises for this chapter.

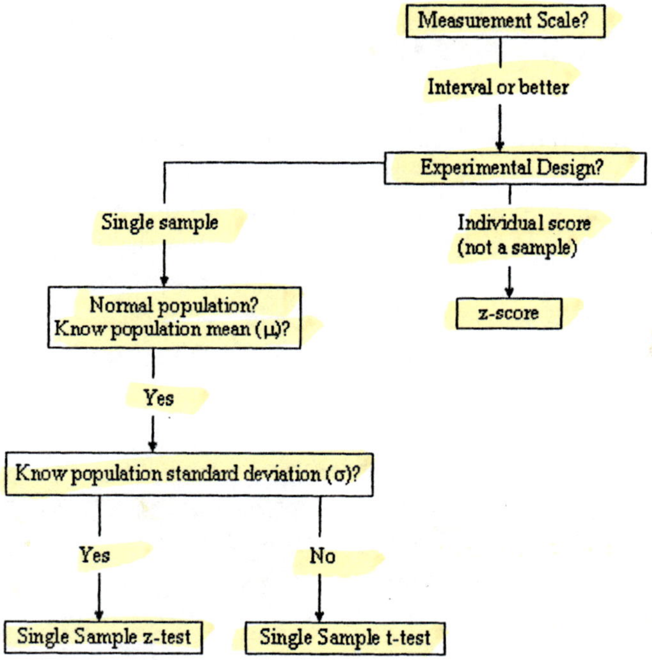

Figure 6.6

Confidence Intervals: A Valuable Use of the Single Sample t test Formula

In everyday life you use **confidence intervals**, but you don't think of them as **confidence intervals**. For example, if your spouse asks you how much money you spend at the race track, you might say, "$50, give or take $5." Essentially you are suggesting that your best guess is $50, but you are not entirely confident that $50 is the exact amount you spent, but you are willing to say that you spent $5 more or less of that amount. Mathematically, you are saying 50 +/- 5, or 45 to 55 dollars. We use **confidence intervals** in the same way in statistics, but we assign probabilities to inform us of how confident we are that the range we have indicated includes the actual number of interest.

We use a rearrangement of the Single Sample t test in order to calculate those probabilities. For example, we might want to be 95% sure that the interval that we describe contains the true value (or population mean, in the terminology of a single sample test). If so, we can calculate a 95% confidence interval that has a probability of 0.95 that the interval that is calculated contains the mean. We are NOT calculating the probability that the mean is in the interval, but rather the probability that the interval contains the mean. This distinction is necessary since the population mean is a fixed, "true" thing, but the confidence interval is what is being estimated.

Confidence Interval for the Population Mean: range of values with a calculated probability of containing the mean.

Confidence Limits for the Population Mean: the upper and lower values (or boundaries) surrounding the confidence interval.

How to we derive the confidence interval from the Single Sample t test Formula?

If you think about the "critical regions of rejection" figures you looked at earlier, determining a mean, plus or minus some value, is similar to that concept. In fact, it would look something like Fig. 6.7 below if you overlaid the t-critical values associated with an alpha of 0.05 and two tails.

Thus, the true value of t-obtained has a 95% chance of being between the t-critical value on the negative tail and the t-critical value in the positive tail.

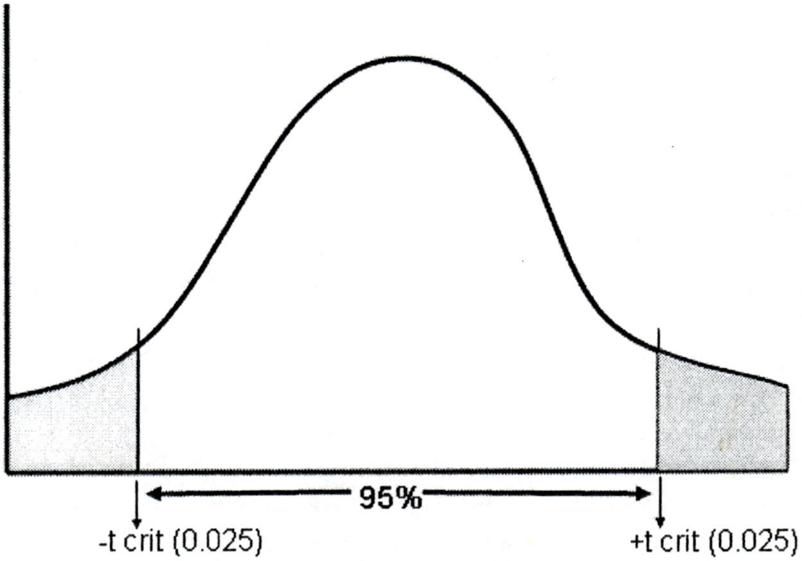

Figure 6.7

Mathematically, that would be $-t_{0.025} \leq t_{obtained} \leq +t_{0.025}$.

However, $t_{obtained} = \dfrac{\overline{X} - \mu}{s_{\overline{X}}}$

If you substitute the formula that we use for t-obtained into the confidence limit formula above, you would have the following:

$$-t_{0.025} \leq \frac{\overline{X} - \mu}{s_{\overline{X}}} \leq +t_{0.025}$$

If you solve for μ, then you get the following:

$$\overline{X} - s_{\overline{X}}\,(t_{0.025}) \leq \mu \leq \overline{X} + s_{\overline{X}}(t_{0.025})$$

This suggests that there is a 95% chance that the interval $\overline{X} \pm s_{\overline{X}}(t_{0.025})$ contains μ. Notice that the statement includes a plus and a minus after the sample mean.

$$\mu_{Lower} = \overline{X} - s_{\overline{X}}(t_{0.025})$$
$$\mu_{Upper} = \overline{X} + s_{\overline{X}}(t_{0.025})$$

This is the same as our earlier statement that you have your best guess (the sample mean) and you add and subtract some amount of error around it to get a confidence interval that you believe contains the true value (or population mean in this case). Let's do an example to clarify any confusion.

Confidence Interval Example

Assume that we have a sample with $\overline{X}$=75, s = 9.37, and N=47. What is the population mean? Our best estimate is, of course, 75. But how sure are we of this estimate? In other words, what is the 95% confidence interval for the population mean, μ?

Step One: Determine the appropriate t-critical values for your α = 0.05 and degrees of freedom (df).

Your df = n-1 = 47-1 = 46. From Table B, you determine that the t-critical values for these degrees of freedom are +/- 2.021.

Step two: Calculate the standard error

$$s_{\overline{X}} = \frac{s}{\sqrt{n}}$$

$$s_{\overline{X}} = \frac{9.37}{\sqrt{47}} = 1.3667$$

Step three: Insert your values into the formula

$\mu_{Lower} = 75 - 1.3667(2.021) = 72.24$

$\mu_{Upper} = 75 + 1.3667(2.021) = 77.76$

Step four: Interpret your confidence interval

There is a 95% probability that the interval from 72.24 to 77.76 contains the population mean. So, our best estimate of the population mean is 75, and we are very sure (95% sure) that the interval between 72.2 and 77.8 contains the population mean. This would be quite different than saying, say, that the mean was 75 and that we were quite sure (95% sure) that the interval between 55 and 95 contained the population mean, a much larger confidence interval indicating much less confidence that we really know what interval contains the population mean. This larger confidence interval would be the result of either a smaller sample size or a larger sample standard deviation, or both.

Usefulness of Confidence Intervals

We have worked through an example involving a 95 % confidence interval, but it possible to calculate confidence intervals of 90% or 99% or any confidence interval that you like. What changes is the t-critical value that you insert into the equation. To determine the confidence interval with 99% confidence means that your confidence interval will actually be larger than the size of a 95% confidence interval. This seems counterintuitive at first, but if you think carefully, you will realize that to be 99% sure that the interval contains the mean, you have make the interval larger. If we had calculated a 99% confidence interval in the example above, the t-critical value would have been 2.704 rather than the 2.021 that we used. Thus, we end up adding a larger number and subtracting a larger number, which increases the size of our interval. A number of statisticians advocate using confidence intervals rather than hypothesis tests to determine the likelihood that a sample mean came from the null hypothesis population. While we recognize the logic behind this approach, our strategy will be to teach you both the confidence interval approach as well as the hypothesis tests since hypothesis testing is still prevalent in the scientific literature.

Summary

In this chapter, we have introduced our first hypothesis tests, those that test hypotheses having to do with single sample designs when compared to underlying populations. We began with a simple application of the concept of standardized z-scores which became the single subject z test. We then introduced designs which compare a sample mean to a population mean, rather than a single score to a population mean, and developed the single subject z test. We went on to make this concept more useful by relaxing the (admittedly uncommonly satisfied) assumption of a known population standard deviation by allowing the estimation of the population standard deviation from the sample standard deviation: the single sample t test. This required a different sampling distribution for calculating the resulting probabilities of the null hypothesis: the t distribution, which is influenced by sample size, or actually degrees of freedom. Finally, we applied the single sample t test to a very useful descriptive statistic called the confidence interval. In the next chapter, we will introduce a new design, two samples, and describe how the t test can be adapted to test hypotheses with this design.

Chapter Six Exercises

Define the following terms.

1. What is the *sampling distribution of a statistic*?

2. What is the *null-hypothesis population*?

3. Why is it called the *null-hypothesis population*?

4. What is the *sampling distribution of the mean*?

5. What is the *Central Limit Theorem*?

Answer these short answer questions.

6. Why is the *standard deviation of the means* also called the *standard error*?

7. If we must estimate the population standard deviation to determine the significance of the sample results, the appropriate inference test is:

8. List three assumptions of the single sample z-test.

Multiple Choice Questions

9. In order to reject the null hypothesis, which of the following decision rules must be true?
   A. $t_{obt} < t_{crit}$
   B. $t_{obt} \geq t_{crit}$
   C. $t_{obt} \leq t_{crit}$
   D. $t_{obt} > t_{crit}$

10. The sampling distribution of t varies with __________
    A. μ
    B. $\bar{X}$
    C. σ
    D. s
    E. df

11. As the df increases, the t distribution ___________
    A. gets larger
    B. gets smaller
    C. stays the same
    D. approaches the normal distribution

12. With which statistical test (s) do we determine the probability of getting the obtained results or those more extreme based on the likelihood of them occurring due to chance alone?
    A. z test
    B. t test
    C. none of the above
    D. a & b

13. If we do not know the population standard deviation, but we do know the population mean and we want to compare it to a sample, the appropriate statistical test is ______
    A. t test
    B. binomial distribution test
    C. z test
    D. all of the above

14. How does the Single sample t-test differ from the Single sample z-test?
    A. with the t test we estimate μ
    B. with the z test we estimate μ
    C. the z distribution is a family of curves
    D. the t distribution is a family of curves
    E. none of the above

15. How are the degrees of freedom for the Single sample t-test determined?
    A. n
    B. n-1
    C. n-2
    D. df-1

16. Due to high publication costs, a social scientist believes that the average number of pages in a professional journal in the social sciences has declined in the last few decades. A previous study indicated that between 1920 and 1970 the average page length of 70 social science journals was 115. To investigate this belief, she randomly samples 41 social science journals and records their page lengths. The journals have a mean of 111 pages and a standard deviation of 12.4 pages. Using $\alpha = .01_{1tail}$, what do you conclude?

17. Assess the following statistics (accept or reject the null hypothesis) under the conditions described. Use the most powerful and appropriate test statistic.
17a. You have a control population of 10,000; $\mu = 67.6$, $\sigma = 5.5$; 1 tailed, $\alpha = 0.01$; experimental sample: $N = 30$, $\bar{x} = 72.1$, $s = 15.6$.

17b. You have a normally distributed population of 35; $\mu = 1.36$, $\sigma = 1.17$; 2 tailed, $\alpha = 0.05$; sample $n = 3$, $\bar{x} = 2.04$, $s = 0.15$.

Choose the most appropriate test to analyze the following story problems. You do not need to calculate the statistic, but merely list the appropriate test for analyzing this scenario.

18. A researcher is interested to see if there is a relationship between aging and depression. On a standardized depression test, the general population averages 40. The researcher tests a random sample of 45 individuals who are all over the age of 70 on the depression test and obtains a mean of 44.5. Does there appear to be a relationship between depression and aging?

19. A health club is interested to find out if individuals under 28 spend more time in the gym than the rest of the general population. The average time spent in the gym per week for the average population is 3.5 hours with a standard deviation of 2.7. The mean hours of time spent in the gym of the 35 individuals under the age of 28 who participated in the study was 5.6 with a standard deviation of 1.8. Test for statistical significance.

Chapter 7
Two-Sample Tests

In the last chapter, you learned how to evaluate hypotheses for tests when you had one sample and known population parameters. While those tests are powerful, population parameters can be difficult to obtain and may be suspect depending on how and when they were obtained. If the parameters are suspect, they can confound statistical interpretation. Here we introduce the two-sample test, where you will compare two samples that came from the same population, rather than comparing a single sample to a population. The samples may be completely independent from one another (between-groups design) or they may be related in some way (within-groups design).

Independent or between-groups designs are those in which subjects are randomly selected from a population and are randomly assigned to either the control or experimental conditions. Subjects only serve in one condition. Dependent or within-groups designs are those in which subjects are randomly selected from a population and serve in more than one condition (such as "before" versus "after" some treatment) or subjects are matched into pairs and one subject in each pair serves in each condition. Because either the same subjects or subjects that are similar to one another in some significant way serve in the with-groups design experiments, the amount of variation due to nuisance factors is minimized in these designs. When the variability that is not of interest to the researchers is minimized, the power of the experiment is increased. Thus, within-groups designs are more powerful than between-groups designs, and we will introduce the test to analyze those designs first.

Paired t-test (Correlated Groups t-test)

The paired t-test (also called the "correlated groups" t-test) is used when you have two samples and a within-groups design. This design is also called a dependent or repeated-groups design. Both the name of the statistical test and the name of the research design can vary a great deal from book to book and between different statistical software packages. You can navigate this confusion by having a conceptual understanding of what the test is doing. This statistical test requires that you have met one of the following experimental design conditions:

1) You have two measures on the same subjects ("before" and "after" measures are common). See the example in Table 7.1. OR
2) You have two separate samples but the subjects in each are individually matched so that there are similar subjects in each group (but not the same subjects in each group). For example, you might match subjects on age and sex, so that you have a 36-year-old female in your control group and a 36-year-old female in your experimental group, a 28-year-old male in your control group and a 28-year-old male in your experimental group, and so on. This can also be done by placing one identical twin in the control group and the other twin in the experimental group, or by any matching of individuals that is an attempt to reduce the variability between the groups. See the example in Table 7.2. Note that the matching must be pair-wise, so that you can literally compare the scores of the twins side by side. You'll see why this is important when you see the formula for the paired t-test.

Table 7.1: An example of "before" and "after" pairing using the same subjects in each "paired" sample

Subject	Score Before Treatment	Score After Treatment
1	50	55
2	52	58
3	44	48
4	42	41
5	49	56

Table 7.2: An example of a "paired" design in which the actual subjects in each sample are different but are "matched" for characteristics which they have in common (in this example, genetics)

Twin Pair	Twin 1 =Control	Twin 2 = Experimental
A	10	8
B	12	10
C	21	19
D	18	15

Paired t-test calculation

The calculation of the paired t-test statistic comes from a modification of the single samples t-test. However, now we first calculate a difference score for each pair of scores in our two samples and treat those difference scores as a single sample that will be compared to the mean difference score (μ_D) of the null hypothesis population. The mean difference score of the null hypothesis population is assumed to be zero ($\mu_D = 0$); that is, no difference between our samples or no effect of our independent variable. The mean difference score for the paired samples is a measure of any effect of our treatment. If our treatment does not have an effect, then there will not be a difference between the two groups, and the mean difference score will be zero (or close to it), like the mean difference score of the null hypothesis population. However, if the treatment does have an effect, it will increase or decrease the scores from the control condition, and therefore produce a mean difference score greater or less than zero.

Table 7.3: An example of difference score calculations for a paired t-test.

Twin Pair	Twin 1 =Control	Twin 2 = Exp	Difference Score
A	10	8	2
B	12	10	2
C	21	19	2
D	18	15	3
			$\sum D = 9$
			$\overline{D} = 9/4 = 2.25$

Thus, we can calculate the sum of the difference scores ($\sum$ D), the sum of the squared difference scores ($\sum D^2$), as well as the mean ($\overline{D}$) and the standard deviation of the difference scores (s_D). These difference scores become our single sample of raw scores that we contrast with a null hypothesis population mean of no difference between subjects ($\mu_D = 0$), and we estimate the standard deviation of the population difference scores (σ_D) based on the sample difference scores, just as we did in the single sample t-test. Because we are using an estimate of the population difference scores based on the sample, but we assume that we know the population mean (=0), we use the t distribution to evaluate our t obtained value.

Formula for a single sample t-test (review)

$$t_{\text{obtained}} = \frac{\overline{X} - \mu}{s_{\overline{X}}} = \frac{\overline{X} - \mu}{\frac{s}{\sqrt{n}}}$$

Paired t-test formula

$$t_{\text{obtained}} = \frac{\overline{D} - \mu_D}{s_{\overline{D}}} = \frac{\overline{D} - \mu_D}{\frac{s_D}{\sqrt{n}}} = \frac{\overline{D} - \mu_D}{\sqrt{\frac{SS_D}{n(n-1)}}}$$

Complete Example

A cognitive psychologist interested in the decay of long-term memory compared the number of errors in memory that an individual made after one week and after one year for a specific event. Subjects viewed a videotape of a bank robbery and were asked a number of specific questions about the video one week after viewing it. They were asked the same questions one year after seeing the video. The number of memory errors was recorded for each subject at each time period. The researchers asked whether or not there was a significant difference in the number of errors in the two time periods. Assume that the difference scores are normally distributed and let $\alpha = 0.05$.

Null Hypothesis: There is no difference in the number of errors made at one week and at one year.

Alternative Hypothesis: There is a difference in the number of errors made at one week and at one year.

Step One: Compute the probability of the mean differences of this sample given that the sample comes from the null hypothesis population of difference scores where $\mu_D = 0$.

Calculate the Difference Scores and the Intermediate Numbers for the SS Formula

Subject	One Week	One Year	Difference Score	D^2
1	5	7	*-2*	*4*
2	4	5	*-1*	*1*
3	6	9	*-3*	*9*
4	8	9	*-1*	*1*
5	6	6	*0*	*0*
6	5	6	*-1*	*1*
7	4	5	*-1*	*1*
8	5	4	*+1*	*1*
9	7	7	*0*	*0*
			$\sum D = -8$	$\sum D^2 = 18$

$n = 9$

$\overline{D} = -8/9 = -0.8888$

Calculate the standard deviation of the sample

$$SS_D = \sum D^2 - \frac{(\sum D)^2}{n} = 18 - \frac{(-8)^2}{9} = 18 - 7.1111 = 10.8888$$

$$s_D = \sqrt{\frac{SS_D}{(n-1)}} = \sqrt{\frac{10.8888}{9-1}} = \sqrt{1.36110} = 1.16666$$

Apply the formula to our example

$$t = \frac{\overline{D} - \mu_D}{\frac{s_D}{\sqrt{n}}}$$

$$t = \frac{-0.8888 - 0}{\frac{1.16666}{\sqrt{9}}} = \frac{-0.8888}{0.388887} = -2.2855$$

Step Two: Evaluating the probability of obtaining this score due to chance

Evaluate the t-obtained value based on alpha (α) = 0.05, and a two-tailed hypothesis. To evaluate your t-obtained value, you must use the t distribution (Table B) as you did in Chapter 6. To determine your t-critical value, you need to know your alpha level (0.05), the number of tails you are evaluating (2, in this case) and your degrees of freedom (n-1 = 9-1 = 8). Compare the t-critical value with your t-obtained value. When α = 0.05, your degrees of freedom are equal to 8, and given that your hypothesis is two tailed, you should use 2.306 as your t-critical value.

To reject the null hypothesis for a two-tailed test, the absolute value of t-obtained must be equal to, or more extreme than, the t-critical value.

$$|t_{obtained}| \geq |t_{critical}|$$

$|-2.2855| \leq |2.306|$, so we *fail-to-reject* the null hypothesis.

How should we interpret these data in light of the effect of time on the number of memory errors?

These results suggest that more than 5% of the time you would obtain this number of memory errors regardless of whether it was after one week or one year. Thus, it is likely that these memory error differences come from the normal null hypothesis population of difference scores. However, remember that there is a chance that there is a real effect of time on memory errors that we have not detected in this analysis.

Results if you use Microsoft Excel to calculate the t-test

t-Test: Paired Two Sample for Means

	One Week	*One Year*
Mean	5.555556	6.444444
Variance	1.777778	3.027778
Observations	9	9
Pearson Correlation	0.742315	
Hypothesized Mean Difference	0	
Df	**8**	
t Stat	**-2.28571**	
P(T<=t) one-tail	0.025804	
t Critical one-tail	1.859548	
P(T<=t) two-tail	0.051609	
t Critical two-tail	**2.306006**	

This is the output that you get from Excel when you type in the data for week one and one year. Note that Excel calls this test "t-Test: Paired Two Sample for Means". This wording is slightly different than we have been using, but it is describing exactly the same analysis. We have bolded the numbers that are comparable to the numbers we just manually calculated or looked up in the table. Note that the t-obtained value (Excel calls it "t Stat") is identical to ours, except they have carried more digits (-2.28571). The "t Critical two-tail" is also the same (2.306006), but again they have carried more digits. The degrees of freedom are the same (8). But they have also provided you with the exact probability of obtaining a t value (t Stat) of -2.28571, which is 0.051609. Since 0.051609 > 0.05, we clearly cannot reject our null hypothesis. Thus, we fail-to-reject the null hypothesis. These results suggest that exactly 5.1609% of the time you would obtain this number of memory errors regardless of whether it was after one week or one year. Knowing the exact probability for each and every t obtained value (not just the t critical values) is one of the major advantages of using a computer to calculate your analyses.

Results if you use SPSS to calculate the t-test
Paired Samples Test (see next table)

We have placed the numbers that are comparable to our manual calculations in bold. Once again, you see that that exact probability (SPSS calls it "Sig. 2-tailed") is greater than our alpha level of 0.05, thus we must assume that these results could occur by chance and not necessarily as a result of time since the event.

		Paired Differences							
		Mean	Std. Deviation	Std. Error Mean	95% Confidence Interval of the Difference		t	df	Sig. (2-tailed)
					Lower	Upper			
Pair 1	ONEWEEK - ONEYEAR	**-.8889**	**1.1667**	**.3889**	-1.7857	7.890E-03	**-2.286**	**8**	.052

Results if you use SPSS to calculate the t-test

Paired Samples Test

How would these results be reported in a scientific journal article?

If the journal required American Psychological Association (APA) format, the results would be reported in a format something like this:

> There was a significant difference in the number of errors made at one week and at one year, $t(8) = -2.29, p > .05$.

This formal sentence includes the dependent variable (number of errors), the independent variable (one week versus one year), as well as a statement about statistical significance, the symbol of the test (t), the degrees of freedom (8), the statistical value (-2.29), and the estimated probability of obtaining this result simply due to chance (>.05).

Another Complete Example

An animal behaviorist is concerned about the effects of nearby construction on the nesting behavior (trips to nest per hour) of endangered dusky seaside sparrows in Florida. She knows that the quality of the nesting territory's habitat will also influence this nesting behavior, so she picks seven pairs of nests, each with the same territory quality (say, density of seed plants), one nest of the pair near the construction and one in an undisturbed location, for a total of 14 nest observations. Assume that the difference scores are normally distributed and let $\alpha = 0.05$.

Null Hypothesis: There is no difference in the rate of nest visits made at "construction" nests and "undisturbed" nests.

Alternative Hypothesis: There is a difference in the rate of nest visits between the two locations. No specific direction of the difference is suggested.

Step One: Compute the probability of the mean differences of this sample given that the sample comes from the null hypothesis population of difference scores where $\mu_D = 0$.

Calculate the Difference Scores and the Intermediate Numbers for the SS Formula

	Rate of Nest Visits per Hour			
Matched Pair	Undisturbed	Construction	Difference Score	D^2
1	5.4	3.2	***2.2***	***4.84***
2	4.1	3.5	***0.6***	***0.36***
3	9.7	7.1	***2.6***	***6.76***
4	8.4	6.8	***1.6***	***2.56***
5	6.0	6.4	***-0.4***	***0.16***
6	6.0	4.5	***1.5***	***2.25***
7	7.9	7.6	***0.3***	***0.09***

$\sum D = 8.4$

$\sum D^2 = 17.02$

$n = 7$

$\overline{D} = 8.4/7 = 1.2$

Calculate the standard deviation of the sample

$$SS_D = \sum D^2 - \frac{(\sum D)^2}{n} = 17.02 - \frac{(8.4)^2}{7} = 17.02 - 10.08 = 6.94$$

$$s_D = \sqrt{\frac{SS_D}{(n-1)}} = \sqrt{\frac{6.94}{7-1}} = \sqrt{1.15667} = 1.07548$$

Apply the formula to our example

$$t = \frac{\overline{D} - \mu_D}{\frac{s_D}{\sqrt{n}}}$$

$$t = \frac{1.2 - 0}{\left(\frac{1.07548}{\sqrt{7}}\right)} = \frac{1.2}{0.40649} = 2.95210$$

Step Two: Evaluating the probability of obtaining this score due to chance

Evaluate the t-obtained value based on alpha (α) = 0.05, and a two-tailed hypothesis. To evaluate your t-obtained value, you must use the t distribution (Table B) as you did in Chapter 6. To determine your t-critical value, you need to know your alpha level (0.05), the number of tails you are evaluating (2, in this case) and your degrees of freedom (n-1 = 7-1 = 6). Compare the t-critical value with your t-obtained value. When $\alpha = 0.05$, your degrees of freedom are equal to 6, and your hypothesis is two tailed, you should use 2.447 as your t-critical value.

To reject the null hypothesis for a two-tailed test, the absolute value of t-obtained must be equal to, or more extreme than, the t-critical value.

$$|t_{obtained}| \geq |t_{critical}|$$

$|2.952| \geq |2.447|$, so we *reject* the null hypothesis.

How should we interpret these data in light of the effect of construction on the rate of nest visits?

These results suggest that <u>less than 5%</u> of the time you would obtain this rate of nest visits regardless of whether it was near or not near to the construction site. Thus, it is likely that the rate of nest visits near the construction site *does not* come from the same underlying population of scores as the nest site visits away from the construction site and therefore, the difference scores in this example do not represent a null hypothesis population of difference scores. However, remember that there is a chance (however small) that there is, in reality, no real effect of the construction on nest site visits, and our conclusion is in error.

Results if you use Microsoft Excel to calculate the t-test
t-Test: Paired Two Sample for Means

	Undisturbed	*Construction*
Mean	6.785714	5.585714
Variance	3.784762	3.284762
Observations	7	7
Pearson Correlation	0.838487	
Hypothesized Mean Difference	0	
Df	**6**	
t Stat	**2.952067**	
P(T<=t) one-tail	0.012772	
t Critical one-tail	1.943181	
P(T<=t) two-tail	0.025544	
t Critical two-tail	**2.446914**	

This is the output that you get from Excel when you type in the data for undisturbed and construction sites. Once again, we have bolded the numbers that are comparable to the numbers we just manually calculated or looked up in the table. Note again that the t-obtained value is identical to ours, except they have carried more digits. The "t Critical two-tail" is also the same, but again they have carried more digits. The degrees of freedom are the same (6). But they have also provided you with the exact probability of obtaining a t value (t Stat) of 2.952067, which is 0.025544. Since 0.025544 < 0.05, we can reject our null hypothesis. Thus, we reject the null hypothesis. These results suggest that exactly 2.5544% of the time you would be making a mistake in concluding that there was a "real effect" of the construction on nest site visits. Traditionally, we are willing to accept a 5% probability of making such a mistake. Again, knowing the exact probability for each t-obtained value (not just the t critical values) is one of the major advantages of using a computer to calculate your analyses.

Results if you use SPSS to calculate the t-test

Paired Samples Test

		Paired Differences Mean	Std. Deviation	Std. Error Mean	95% Confidence Interval of the Difference Lower	Upper	t	df	Sig. (2-tailed)
Pair 1	UNDISTUR - CONSTRUC	**1.2000**	**1.0755**	**.4065**	.2053	2.1947	**2.952**	**6**	.026

Once again, we have placed the numbers that are comparable to our manual calculations in bold. And once again, you see that exact probability (SPSS calls Sig. 2-tailed) is greater than our alpha level of 0.05, thus we must assume that these results could occur by chance and not necessarily as a result of time since the event.

How would these results be reported in a scientific journal article?

If the journal required American Psychological Association (APA) format, the results would be reported in a format something like this:

> There is a significant difference in the rate of nest visits between the two locations, $t(6) = 2.95, p = .03$.

This formal sentence includes the dependent variable (rate of nest visits), the independent variable (two different locations), as well as a statement about statistical significance, the symbol of the test (t), the degrees of freedom (6), the statistical value (2.95), and the estimated probability of obtaining this result simply due to chance (.03).

When is it appropriate to use a paired t test?

1) Use a paired t test when you have two samples and a within groups design, and
2) When the sampling distribution is normally distributed, which, as you should recall, is satisfied when:
    c. The sample size is greater or equal to 30 ($N \geq 30$) or,
    d. The null hypothesis population is known to be normally distributed.

   and

3) When the dependent variable is on an interval or ratio scale.

Independent t-test

In the previous section we described a situation where your two conditions contain either the same subjects or subjects that have been individually matched on an important characteristic that might potentially influence the outcome of your results and is not interesting to you (age or body weight are potential examples of characteristics that you might match subjects on). Now we turn to the situation where you actually have two completely different (independent) groups of subjects that you want to compare to determine if they are significantly different from one another. The classic example of this is when you have a sample and you randomly assign half of your subjects to the control condition and the other half to the experimental treatment condition. In this situation we wish to compare the means of the two conditions/groups. We can no longer assume that we know a population mean (even one where $\mu_D = 0$), and we must develop a new sampling distribution.

Sampling distribution of the difference between the means

In order to test for the potential statistical significance of a true difference between sample means, we need a sampling distribution of the difference between sample means ($\overline{X}_1 - \overline{X}_2$). Thus, what we really need is a sampling distribution that will tell us the probability of getting two sample means ($\overline{X}_1, \overline{X}_2$) that differ from the sample null hypothesis population of sample means. The sampling distribution of the difference between the means can be created by taking all possible sample sizes of n_1 and n_2, calculating the sample means, and then taking the difference of those means. If you do this repeatedly for all of the possible combinations of your sample sizes, then you end up with a distribution of differences between the two means when they are randomly drawn from the same null hypothesis population.

The sampling distribution of the difference between sample means has the following characteristics:

1) If the null hypothesis population of scores are normally distributed, then the population of differences between the sample means will also be normally distributed, and
2) The mean of the sampling distribution of the difference between sample means ($\mu_{\overline{X}_1 - \overline{X}_2}$) will be equal to $\mu_1 - \mu_2$ (just as $\mu_{\overline{X}} = \mu$), and

3) The standard deviation of the sampling distribution of the difference between sample means will be equal to the square root of the sum of each sample variance, or $\sqrt{\sigma_{\overline{X}_1}^{\ 2} + \sigma_{\overline{X}_2}^{\ 2}}$.

Estimating Variance from Two Samples

Consider for a moment that the formulas we've used in the past contained one measure of variability . In the case of the single sample z-test we used the population standard deviation, while with the single sample t-test we estimated the population standard deviation with the sample standard deviation. When we performed the paired t-test earlier in this chapter, we calculated the standard deviation of the sample of difference scores, so we were able to come up with one measure of sample variability. Now we have two independent samples and each has a sample standard deviation. This forces us to determine which standard deviation best represents the true variability. In fact, what we do is to estimate the true population variability (or variance, σ^2) by taking the average variance (s^2) of our samples, weighted by their respective sample sizes. Remember, as we learned in earlier chapters, sample size or degrees of freedom affect the accuracy of our variance estimates, so an estimate from a sample with a large sample size would be more accurate than an estimated variance from a smaller sample. So we need to weight our average variance by the respective sample sizes of each sample. In using this approach, we are going to make a new assumption: that the sample variances are estimating the same underlying population variance, the variance of the null hypothesis population. Later in this chapter, we will have to check to make sure that our two sample variances are the same, within the bounds of random sampling error.

Formula for Weighted Variance

$$s_W^{\ 2} = \frac{df_1 s_1^{\ 2} + df_2 s_2^{\ 2}}{df_1 + df_2}$$

Substituting the equation for degrees of freedom and for variance:

$$= \frac{(n_1 - 1)(SS_1/n_1 - 1) + (n_2 - 1)(SS_2/n_2 - 1)}{(n_1 - 1) + (n_2 - 1)}$$

Rearranging to simplify:

$$= \frac{SS_1 + SS_2}{n_1 + n_2 - 2}$$

We have shown the formula in three different ways. The first way is the most intuitive way to present the average variance of the two samples when it is weighted by the sample size, or more specifically, by the appropriate degrees of freedom for each sample. The degrees of freedom are used because we are estimating the population variation from a sample, and one degree of freedom is lost each time we do that (one for each sample). The second formula actually plugs in the appropriate formulas for variance and degrees of freedom into the first formula, and the last formula is created by algebraic rearrangement into a simplified version. The first formula may be the best one for you to use if you are obtaining each sample variance from your calculator directly from the raw data, or if you are given either the variance or standard deviation of each sample in a problem. The last formula would be best if you have already calculated the Sums-of-squares (SS) for each group.

Derivation of the Independent t-test formula from the Single Sample t-test formula

Formula for a Single Sample t-test (for review):

$$t_{\text{obtained}} = \frac{\overline{X} - \mu}{\frac{s}{\sqrt{N}}}$$

Conceptual Formula for an Independent t-test:

$$t_{\text{obtained}} = \frac{(\overline{X}_1 - \overline{X}_2) - \mu_{\overline{X}_1 - \overline{X}_2}}{\sqrt{\frac{\sigma_1^2}{n_1} + \frac{\sigma_2^2}{n_2}}} = \frac{(\overline{X}_1 - \overline{X}_2) - \mu_{\overline{X}_1 - \overline{X}_2}}{\sqrt{\sigma^2 (1/n_1 + 1/n_2)}}$$

Calculation (practical) Formula to use for an Independent t-test:

$$t_{obtained} = \frac{\overline{X}_1 - \overline{X}_2}{\sqrt{s_w^{\,2}(1/n_1 + 1/n_2)}}$$

The formula for an independent t-test is derived by assuming the mean of the sampling distribution or differences between means is zero for the null hypothesis population, and by using the average variance divided by each sample size. The square root in the denominator of the independent t-test formula is there not only to take the square root of the sample size (as you did when you calculated a single sample t-test), but also because you are working in squared units (variance) and we must take the square root of the variance to get back to the standard deviation.

Variations of the Independent t-test formula

$$t_{obtained} = \frac{\overline{X}_1 - \overline{X}_2}{\sqrt{s_w^{\,2}(1/n_1 + 1/n_2)}} = \frac{\overline{X}_1 - \overline{X}_2}{\sqrt{\left(\frac{SS_1 + SS_2}{n_1 + n_2 - 2}\right)(1/n_1 + 1/n_2)}}$$

To create the second variation of the formula, we simply substituted one of the formulas for $s_w^{\,2}$ directly into the independent t-test formula.

1) **Complete Example**

A researcher breeds rats for nine generations, but only breeds the rats that perform very well in a maze (few errors) to each other (called "maze bright" rats), and also breeds rats that perform very poorly in a maze (many errors) to one another ("maze dull" rats). After nine generations, is there a significant difference in the number of errors in the two groups of rats?

Null Hypothesis: There is no difference in the number of errors made by the maze bright and the maze dull rats or the differences are due to chance.

Alternative Hypothesis: There is a difference in the number of errors made by each group.

Step One: Compute the probability that each of the sample means comes from the null hypothesis population of differences between means. This can be done using the weighted variance formula (uses descriptive statistics) or the sums-of-squares (SS) formula (from raw data). We will demonstrate the steps for the same data set calculated using both techniques.

Number of Errors in a Maze by 9th Generation Rats

Maze Bright Rats	Maze Dull Rats
2	4
3	5
4	6
3	3
4	6
2	

Intermediate Steps that can be obtained from your calculator

$\Sigma X_{BRIGHT} = 18$

$(\Sigma X_{BRIGHT})^2 = 18^2 = 324$

$\Sigma X^2_{BRIGHT} = 58$

$\overline{X}_{BRIGHT} = \frac{18}{6} = 3$

$s_{BRIGHT} = .89442$

$n_{BRIGHT} = 6$

$\Sigma X_{DULL} = 24$

$(\Sigma X_{DULL})^2 = 24^2 = 576$

$\Sigma X^2_{DULL} = 122$

$\overline{X}_{DULL} = \frac{24}{5} = 4.8$

$s_{DULL} = 1.30384$

$n_{DULL} = 5$

Solve for s_w^2

$$s_w^2 = \frac{df_1 s_1^2 + df_2 s_2^2}{df_1 + df_2}$$

$df = n-1$

df_1 (BRIGHT) = 5

df_2 (DULL) = 4

$$s_w^{\,2} = \frac{df_1 s_1^{\,2} + df_2 s_2^{\,2}}{df_1 + df_2} = \frac{(5)(.89442^2) + (4)(1.30384^2)}{5+4}$$

$$= \frac{(5)(.799) + (4)(1.7)}{5+4} = \frac{10.795}{9} = 1.1994$$

Calculate your independent t-test using the s_w^2 formula

$$t_{obtained} = \frac{\overline{X}_1 - \overline{X}_2}{\sqrt{s_w^{\,2}(1/n_1 + 1/n_2)}}$$

$$t = \frac{3 - 4.8}{\sqrt{(1.1994)(1/6 + 1/5)}} = \frac{-1.8}{\sqrt{(1.1994)(.3666)}} = -2.714$$

Same Example, but Calculate the SS of each sample

$$SS_{BRIGHT} = \sum X^2 - \frac{(\sum X)^2}{n} = 58 - \frac{(18)^2}{6} = 58 - \frac{324}{6} = 4$$

$$SS_{DULL} = \sum X^2 - \frac{(\sum X)^2}{n} = 122 - \frac{(24)^2}{5} = 122 - \frac{576}{5} = 6.8$$

Apply the SS formula of the independent t-test to our example

$$t_{obtained} = \frac{\overline{X}_1 - \overline{X}_2}{\sqrt{\left(\frac{SS_1 + SS_2}{n_1 + n_2 - 2}\right)(1/n_1 + 1/n_2)}}$$

$$t = \frac{3 - 4.8}{\sqrt{\left(\frac{4 + 6.8}{6 + 5 - 2}\right)(1/6 + 1/5)}} = \frac{-1.8}{\sqrt{(1.2)(.3666)}} = -2.714$$

Step Two: Evaluating the probability of obtaining this score due to chance

Evaluate the t-obtained value based on alpha (α) = 0.05, and a two-tailed hypothesis. To evaluate your t-obtained value, you must use the t distribution (Table B). To determine your t-critical value, you need to know your alpha level (0.05), the number of tails you are evaluating (2 in this case) and your degrees of freedom (df). The degrees of freedom for an independent t-test are $(n_1-1) + (n_2-1)$ or $n_1+n_2 - 2$. Thus, the df for this problem are $6+5-2 = 9$. Compare the t-critical value with your t-obtained value. When $\alpha = 0.05$, your degrees of freedom are equal to 9, and your hypothesis is two tailed, you should use 2.262 as your t-critical value.

$|-2.714| \geq |2.262|$, so we *reject* the null hypothesis.

How should we interpret these data in light of the effect of breeding on the number of memory errors?

These results suggest that less than 5% of the time, you would obtain this difference in the number of errors if the breeding had no effect. Thus, it is not very likely that these error differences come from the normal null hypothesis population. However, there is a chance that you could get a difference this large between two means that is purely due to chance, but that chance is less than 5%.

Results if you use Microsoft Excel to calculate the t-test
t-Test: Two-Sample Assuming Equal Variances

	Bright	*Dull*
Mean	3	4.8
Variance	0.8	1.7
Observations	6	5
Pooled Variance	1.2	
Hypothesized Mean Difference	0	
Df	**9**	
t Stat	**-2.713602101**	
P(T<=t) one-tail	0.011928192	
t Critical one-tail	1.833113856	
P(T<=t) two-tail	0.023856384	
t Critical two-tail	**2.262158887**	

This is the output you get from Excel when you type in the data for maze bright and maze dull rats. Note that Excel calls this test "t-Test: Two-Sample Assuming Equal Variances". This wording is slightly different than we have been using, but it is describing exactly the same analysis. This will become even clearer after we have discussed the assumptions of the independent t-test. We have bolded the numbers that are comparable to the numbers we just manually calculated or looked up in the table. Note that the t-obtained value (Excel calls it "t Stat") is identical to ours, except they have carried more digits (-2.713602101). The "t Critical two-tail" is also the same (2.262158887), but again they have carried more digits. The degrees of freedom are the same (9). In addition, they have provided you with the exact probability of obtaining a t value (t Stat) of -2.713602101, which is 0.023856384. Since 0.023856384 < 0.05, we clearly *reject* our null hypothesis. These results suggest that exactly 2.3856384% of the time you would obtain this difference in errors if breeding had no effect. Again, knowing the exact probability for each and every t-obtained value (not just the t-critical values) is one of the major advantages of using a computer to calculate your analyses.

Results if you use SPSS to calculate the t-test

Independent Samples Test

		Levene's Test for Equality of Variances		t-test for Equality of Means						
		F	Sig.	t	df	Sig. (2-tailed)	Mean Difference	Std. Error Difference	95% Confidence Interval of the Difference	
									Lower	Upper
ERRORS	Equal variances assumed	1.255	.292	**-2.714**	**9**	.024	**-1.8000**	.6633	-3.3005	-.2995
	Equal variances not assumed			-2.616	6.903	.035	-1.8000	.6880	-3.4315	-.1685

We have placed the numbers that are comparable to our manual calculations in bold. Once again, you see that that exact probability (SPSS calls it Sig. 2-tailed) is less than our alpha level of 0.05, thus we must assume that these results would be *unlikely* simply due to chance.

How would these results be reported in a scientific journal article?

If the journal required American Psychological Association (APA) format, the results would be reported in a format something like this:

> There is a significant difference in the number of errors made by maze-dull and maze-bright rats, $t(9) = -2.71, p = .02$.

This formal sentence includes the dependent variable (number of errors), the independent variable (two groups of rats), as well as a statement about statistical significance, the symbol of the test (t), the degrees of freedom (9), the statistical value (-2.71), and the estimated probability of obtaining this result simply due to chance (.02).

Another Complete Example

Psychology experiments frequently need to be very careful about the effects of past experience on the outcome of their studies. One common population for psychology research is college students enrolled in psychology courses. But it is traditional to use only freshmen or sophomores who are enrolled in introductory courses so that the researcher avoids the possibility of a student anticipating the study's goals after having had more advanced psychology courses. But, does classroom experience in psychology really affect the outcome of the experiment? A researcher develops a simple assessment for response to visual stimuli and the number of correct (matching) results is scored for a group of five sophomore introductory psychology course students and six psychology seniors. Is there a significant difference in the scores of the two groups of students?

Null Hypothesis: There is no difference in the scores of introductory and advanced psychology students or the differences are due to chance.

Alternative Hypothesis: There is a difference in the scores of each group.

Step One: Compute the probability that each of the sample means comes from the null hypothesis population of differences between means.

Calculate the Means and the Intermediate Numbers for the SS Formula

Scores for Introductory and Advanced Students

Introductory	Advanced
9	8
8	9
7	8
9	6
6	10
	9

$\Sigma X_{Intro} = 39$ $\Sigma X_{Adv} = 50$

$(\Sigma X_{Intro})^2 = 39^2 = 1521$

$(\Sigma X_{Adv})^2 = 50^2 = 2500$

$\Sigma X^2_{Intro} = 311$ $\Sigma X^2_{Adv} = 426$

$\overline{X}_{Intro} = \frac{39}{5} = 7.80$ $\overline{X}_{Adv} = \frac{50}{6} = 8.33$

Calculate the SS of each sample

$$SS_{Intro} = \sum X^2 - \frac{(\sum X)^2}{n} = 311 - \frac{(39)^2}{5} = 311 - \frac{1521}{5} = 6.80$$

$$SS_{Adv} = \sum X^2 - \frac{(\sum X)^2}{n} = 426 - \frac{(50)^2}{6} = 426 - \frac{2500}{6} = 9.33$$

Apply the SS formula of the independent t-test to our example

$$t_{obtained} = \frac{\overline{X}_1 - \overline{X}_2}{\sqrt{\left(\frac{SS_1 + SS_2}{n_1 + n_2 - 2}\right)(1/n_1 + 1/n_2)}}$$

$$t = \frac{7.80 - 8.33}{\sqrt{\left(\frac{6.80 + 9.33}{5 + 6 - 2}\right)(1/5 + 1/6)}} = \frac{-0.53}{\sqrt{(1.79222)(.36667)}} = -0.6538$$

Step Two: Evaluating the probability of obtaining this score due to chance

Evaluate the t-obtained value based on alpha (α) = 0.05, and a two-tailed hypothesis. To evaluate your t-obtained value, you must use the t distribution (Table B). To determine your t-critical value, you need to know your alpha level (0.05), the number of tails you are evaluating (2 in this case) and your degrees of freedom (df). The degrees of freedom for an independent t-test are $(n_1-1) + (n_2-1)$ or $n_1+n_2 - 2$. Thus, the df for this problem are 5+6 – 2 = 9. Compare the t-critical value with your t-obtained value. When $\alpha = 0.05$, your degrees of freedom are equal to 9, and your hypothesis is two tailed, you should use 2.262 as your t-critical value.

$|-0.6538| \leq |2.262|$, so we *fail-to-reject* the null hypothesis.

How should we interpret these data in light of the effect of time on the number of memory errors?

These results suggest that greater than 5% of the time, you would obtain this difference in the scores if experience level had no effect. Thus, it is likely that these differences come from the normal null hypothesis population. However, there is always a chance that there is a significant real effect that you did not detect.

Results if you use Microsoft Excel to calculate the t-test

t-Test: Two-Sample Assuming Equal Variances

	Introductory	*Advanced*
Mean	**7.800000**	**8.333333**
Variance	1.700000	1.866667
Observations	**5**	**6**
Pooled Variance	1.792593	
Hypothesized Mean Difference	0	
Df	**9**	
t Stat	**-0.657843**	
P(T<=t) one-tail	0.263552	
t Critical one-tail	1.833114	
P(T<=t) two-tail	0.527105	
t Critical two-tail	**2.262159**	

This is the output you get from Excel when you type in the data for introductory and advanced psychology students. We have again bolded the numbers that are comparable to the numbers we just manually calculated or looked up in the table. Note again that the t-obtained value is identical to ours, except they have carried more digits. The "t Critical two-tail" is also the same (2.262158887), and the same as in the first example since the degrees of freedom are the same (9). In addition, they have provided you with the exact probability of obtaining a t value (t Stat) of -0.657843, which is 0.527105. Since 0.527105> 0.05, we clearly *fail-to-reject* our null hypothesis. These results suggest that exactly 52.7105% of the time you would obtain this difference in errors if psychology experience had no effect.

Results if you use SPSS to calculate the t-test

Independent Samples Test

		Levene's Test for Equality of Variances		t-test for Equality of Means						
		F	Sig.	t	Df	Sig. (2-tailed)	Mean Difference	Std. Error Difference	95% Confidence Interval of the Difference	
									Lower	Upper
DATA	Equal variances assumed	.008	.929	**-.658**	**9**	.527	-.5333	.8107	-2.3673	1.3007
	Equal variances not assumed			-.661	8.785	.526	-.5333	.8069	-2.3655	1.2989

We have placed the numbers that are comparable to our manual calculations in bold. Once again, you see that that exact probability (SPSS calls it Sig. 2-tailed) is greater than (*much* greater than) our alpha level of 0.05, thus we must assume that these results would be *likely* simply due to chance.

How would these results be reported in a scientific journal article?

If the journal required American Psychological Association (APA) format, the results would be reported in a format something like this:

> There is no significant difference between the scores of introductory and advanced psychology students, $t(9) = -.66$, $p = .53$.

This formal sentence includes the dependent variable (scores), the independent variable (introductory versus advanced psychology students), as well as a statement about statistical significance, the symbol of the test (t), the degrees of freedom (9), the statistical value (-.66), and the probability of obtaining this result simply due to chance (.53).

When is it appropriate to use an independent t test?

1) Use an independent t-test when you have two samples and a between groups design, and
2) When the sampling distribution is normally distributed. Again, this is satisfied when:
   a. The sample size is greater or equal to 30 ($N \geq 30$) or,
   b. The null hypothesis population is known to be normally distributed.

   and
3) When the dependent variable is on an interval or ratio scale, and
4) When the variances of the two groups are the same, or are **homogeneous**. The homogeneity of variance assumption (HOV) requires that the variances of the two samples are equal, or in

practical terms, not significantly different from one another. For now, we will use a simple rule of thumb to decide whether the variances are similar enough to be considered homogeneous:

Rule of Thumb for HOV check: if the $\dfrac{\textit{Larger Variance}}{\textit{Smaller Variance}} \leq 4$

then HOV assumption is met (okay) and you can proceed with the independent t-test. If not, you cannot proceed with the independent t-test.

Power and Two-sample tests: Paired vs. Independent Designs

Because high sample variability decreases power, and more variability would be expected from between-groups designs where different individuals are in each condition or group, the independent t-test is less powerful than the paired t-test. The paired t-test gains power by controlling nuisance factors such as the age and temperament of a subject since the subjects are either the same in both conditions or they are match on characteristics that might influence the outcome of the study, but are not of interest to the researcher. Thus the researcher is more likely to control variability in the within-groups design.

There is a major drawback to the within-groups (or paired) design, and that occurs when subjects drop out of the experiment prior to experiencing all of the conditions. This creates a situation where the researcher has one data point for the subject but lacks the other data point, and thus cannot calculate a difference score for the subject. There are two options in this situation. The first option is to drop those subjects with missing data from your analysis, but this reduces your power significantly if you have a large number of missing data points relative to your sample size. The second option is to treat the two samples as independent and calculate an independent t-test, thus equal n's are not required and difference scores are not the basis for the test.

Overview of Single Sample and Two Sample Tests: Expanded Flowchart

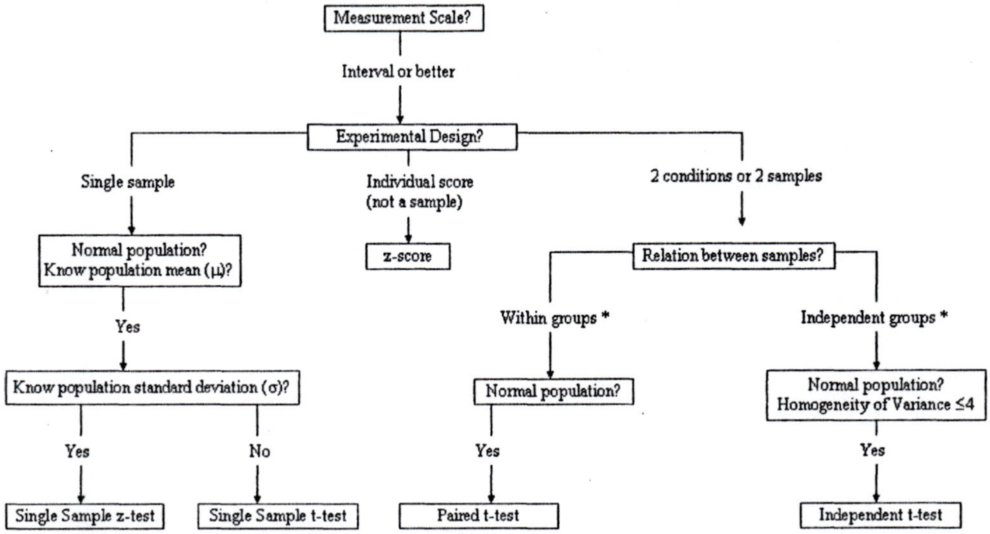

1. Within Groups
   a. subjects matched - married, partners, twins, etc., or individually matched for a reason that might affect the I.V.
   b. before vs. after conditions with the same subjects, measured both times
2. Independent Groups
   a. In one group/condition, NOT allowed in the other, subjects randomly assigned to be in the experiment in one group OR the other
   b. Sample sizes can be unequal (IF unequal, definitely independent t-test, but if equal, have to make decision based on something else).
3. Power
   a. Single z > Single t, Paired > Independent

Notice that the flowchart does not yet tell you what to do if you do not meet the assumptions of your statistical tests. In fact, there are some options when you have two samples, but you do not meet the assumptions for parametric tests. The paired t-test can be replaced with a non-parametric test called the Wilcoxon Signed Ranks Test while the independent t-test can be replaced with a non-parametric test called the Mann-Whitney U test. Another option is to perform a log transformation on your data to determine if the mathematical transformation will result in a normal distribution.

Summary

In this chapter you have covered statistical tests for designs in which you have two samples and you lack information about the underlying population. Because we rarely know population parameters such as the mean and standard deviation (μ and σ), the two sample tests tend to be used more often than the one sample tests, but they are also less powerful. They are less powerful because we are forced to estimate characteristics of the population whereas one sample tests rely on known measures of the population. In the next chapter we will extend the concept of an independent t-test or two-sample between groups design, to a situation where we have three samples and a between groups design.

Chapter Seven Homework

Provide a short answer for the following questions.

1. Why are paired (or correlated) designs more powerful than independent designs?

2. What are the assumptions for the paired t-test?

3. What are the assumptions for the independent t-test?

4. List three ways that you can meet the assumption of normality.

5. Why does the independent t-test require the assumption of homogeneity of variance but the paired t-test does not require this assumption?

6. How should you handle a situation where you have paired design and two conditions, but there is some missing data for one of your two conditions?

7. What do you do if you do not meet the assumptions of the paired t-test or the independent t-test?

8. What is the *Central Limit Theorem*?

9. A cognitive psychologist records the number of positive early childhood memories that can be recalled by five individuals that grew up in military families to the number of memories of individuals that grew up in non-military families. The number of memories is normally distributed in each group.

Military Family	18	25	17	20	23
Non-Military Family	20	23	26	30	28

Using $\alpha=0.05_{2tail}$, what do you conclude?

10. A social psychologist is interested in whether or not race affects the likelihood that the average person will "shoot" a potential criminal in a computer simulation. Subjects are required to make quick decisions about whether to "shoot" or not, and they are shown a variety of images of people. Some of the images are of people with a weapon and some of them are people holding non-violent objects. Eight subjects are randomly sampled for the study. The psychologist records the number of errors (shooting someone holding a non-violent object) the subjects made based on race (African American or Caucasian). The number of errors is normally distributed. The following data are recorded.

Subject	1	2	3	4	5	6	7	8
African-American	28	29	25	30	25	27	28	24
Caucasian	25	28	22	30	26	24	25	22

Using $\alpha = 0.05_{2\text{tail}}$, what do you conclude?

11. Professor Jones is intensely curious about differences in testing situations, and wondered if students tended to make better scores on her tests depending whether the test was taken on a Monday morning or a Friday morning. Her exams have always been normally distributed. From a group of 19 similarly talented students, she randomly selected some to take a test on Friday and others to take it on Monday. The scores by groups were:

Monday	**Friday**
89.8	87.3
90.2	87.6
98.1	87.3
91.2	91.8
88.9	86.4
90.3	86.4
99.2	93.1
94.0	89.2
88.7	90.1
83.9	

Using $\alpha=0.05_{2tail}$, what do you conclude?

Choose the most appropriate and powerful test. Support your answers and list assumptions you are making. Do not try to solve.

12. Extensive research has been done on the subject of birth order. Data on this research shows that first-born children develop different characteristics than later born children. For example, first-born children tend to be more responsible and self-disciplined than later born children. A researcher is interested in finding out if first-born children tend to be more confident and have higher self-esteem than later born siblings. A random sample of 31 first-born children and 35 later-born children were given a self-esteem test. The standard deviation for the first-born is 1.34 and the standard deviation for the later born is 2.15. Test whether birth order affects self-esteem.

13. A researcher tested a new medicine to see if it would be effective in lowering blood pressure. Two samples of subjects from a normally distributed population were matched for medical history and initial blood pressure readings. 15 randomly selected subjects were run through the experimental condition in which they received the new drug. The other 15 randomly selected subjects participated in the control group and received a placebo. Subjects receiving the drug showed a lower blood pressure. Test whether the drug had a statistically significant effect on blood pressure. Variances are homogeneous.

14. A researcher is interested to see if there is a relationship between aging and depression. On a standardized depression test, the general population averages 40. The researcher tests a random sample of 45 individuals who are all over the age of 70 on the depression test and obtains a mean of 44.5. Does there appear to be a relationship between depression and aging?

15. A health club is interested to find out if individuals under 28 spend more time in the gym than the rest of the general population. The average time spent in the gym per week for the average population is 3.5 hours with a standard deviation of 2.7. The mean hours of time spent in the gym of the 35 individuals under the age of 28 who participated in the study was 5.6 with a standard deviation of 1.8. Test for statistical significance.

Chapter 8
Analysis of Variance (ANOVA)

In Unit 3, you will learn to analyze data from complex experimental designs. In this chapter, we address the appropriate statistical test for situations in which you have three or more groups or conditions. This is a commonly used design in behavioral sciences because you can test for the effect of varying degrees of an independent variable. For example, you could test for the effects of caffeine on attention span by having a high-caffeine, low-caffeine and no-caffeine conditions. Alternatively, you could compare 4-year-olds, 8-year-olds and 12-year-olds on their ability to use memory strategies. In this case, your independent variable is age and you are using three levels of age in your experiment. For now, assume your groups are independent of one another, so subjects only serve in one of the three conditions (also called a between-groups design). It is possible to have three or more groups participating in a within-groups (or repeated measures) design, but we will discuss that situation in the next chapter.

Why do we need the ANOVA?

The Analysis of Variance (ANOVA) is similar to the independent t-test discussed in chapter seven, except that you have more than two groups. You may be wondering why you can't simply use the independent t-test to analyze your data. Let's think about the number of tests you would need to perform to determine whether there were any significant differences between one or more of your groups or conditions. What if you had the following results from the experiment described above:

Low Caffeine	High Caffeine	No Caffeine (Control)
$\overline{X} = 20.88$	$\overline{X} = 13.15$	$\overline{X} = 30.29$

You would need to conduct three independent t-tests to determine whether or not there were significant differences amongst these means. To be thorough, you would conduct the following comparisons:
Low Caffeine versus High Caffeine
Low Caffeine versus No Caffeine
High Caffeine versus No Caffeine

What's wrong with numerous pairwise comparisons?

When you perform multiple pairwise comparisons on the same data set, you run the risk of an inflated type I error rate. Recall that we typically limit our type I error rate by setting $\alpha = 0.05$. That means that we are willing to incorrectly reject the null hypothesis 5% of the time. However, when we make so many comparisons on the same data, we are making multiple "draws" from the same distribution. Our probability that we will be able to reject the null hypothesis for one of those comparisons increases, and thus our type I error rate increases. One (bad) solution is to divide our α level by the number of comparisons we wish to make. In the caffeine example that would be $\frac{\alpha}{3}$, or $\frac{.05}{3} = 0.016666$. This is a very conservative approach to solving the type I error rate for pairwise comparisons. If you had five groups/conditions in your study, then you would have nine possible comparisons you could make amongst your means. If you applied this same "correction" to your α level, then you would be forced to use an α level of 0.0055 because $\frac{\alpha}{9}$, or $\frac{.05}{9} = 0.0055$. Obviously, it would be difficult to obtain differences amongst your means large enough to demonstrate statistical significance. Fortunately, there is a better solution to the problem of controlling type I error rate without inflating type II error rate, as you are likely to do when you divide your α level by the number of comparisons you wish to make.

The ANOVA Solution

With the ANOVA, you perform one overall test to determine whether or not one or more of the conditions has different effects than the other conditions. Then, only if the overall test is statistically significant do you

proceed to do pairwise comparisons to determine exactly which comparisons are different from one another.
Your alternative hypothesis (H_A) is that one or more conditions have an effect (one or more means differ from other means), and your null hypothesis (H_0) is that there is no effect (your means are equal). Mathematically, the null hypothesis would be represented in the following way:

$$\mu_1 = \mu_2 = \mu_3 = \mu_k$$

where k = number of groups/conditions.

You must assume that only the means are affected by your experimental manipulations, and that the variance remains unaffected. Mathematically, this is written as:

$$\sigma_1^2 = \sigma_2^2 = \sigma_3^2 = \sigma_k^2$$

Again, "k" represents the number of groups/conditions.

Specifically, we look at the variability between (or among) the group means. We assume this variability is due to our experimental manipulations, and some natural inherent variation. We then compare it to the variability within the groups. Within-group variability should represent the inherent variation in the population. This can be more easily shown graphically. To show you how it works we will first present a figure demonstrating this concept with the independent t-test (two means) and then we can expand the demonstration to include three means.

Figure 8.1

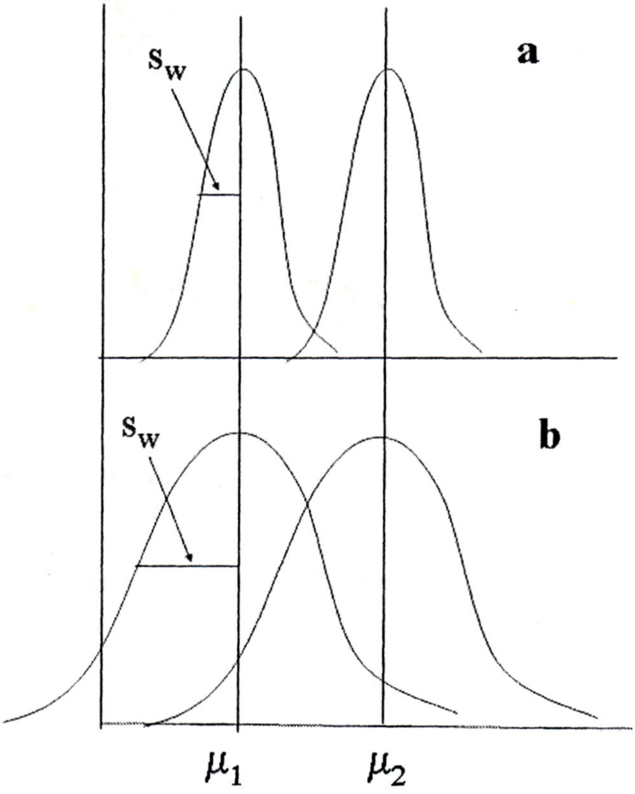

s_W in this figure, as in the independent t-test from the last chapter, denotes pooled estimate of variance from all samples. Note that the means in both examples are the same, but the variances are not the same. There is more variability in example *b* than you see in *a*, but the two groups in example *a* have equal variances, as do the two groups in example *b*. If both groups have the same variance (regardless of whether it is large or small), then you have met the assumption of homogeneity of variance. This means that you can now turn to analyzing the differences between the means, relative to the variability within the groups. This is essentially what you did when you performed the independent t-test calculations in the last chapter, but you may not have thought about it in this way. If we take a moment to recall the formula for the independent t-test, you'll see that we analyze both the differences between the means ($\overline{X}_1 - \overline{X}_2$) as well as some measure of variability, $\sqrt{s_w^2(1/n_1 + 1/n_2)}$.

Formula for an independent t-test

$$t_{obtained} = \frac{\overline{X}_1 - \overline{X}_2}{\sqrt{s_w^2(1/n_1 + 1/n_2)}}$$

The larger the numerator ($\overline{X}_1 - \overline{X}_2$), or the larger the difference between your two sample means, and the smaller the denominator (your measure of variability), then the larger your t-obtained value will become. Of course, when you have a large t-obtained value, you are more likely to be able to reject the null hypothesis. Figure 8.1 shows two examples where the means are equal, but the variance is much higher in example b compared to example a. In this scenario, you can be sure that the t-obtained value for example b will be much smaller than it will be for example a. Thus, even when we were using the independent t-test, the variability (or within groups variance, s_w^2) played a role in determining the outcome of the analyses. When "within group variance" is small relative to the difference between the means, we say there is a significant difference between our means. That continues to be true whether you have two independent (between groups) samples or more than two samples, as is the case with the independent groups (between groups) ANOVA.

Figure 8.2

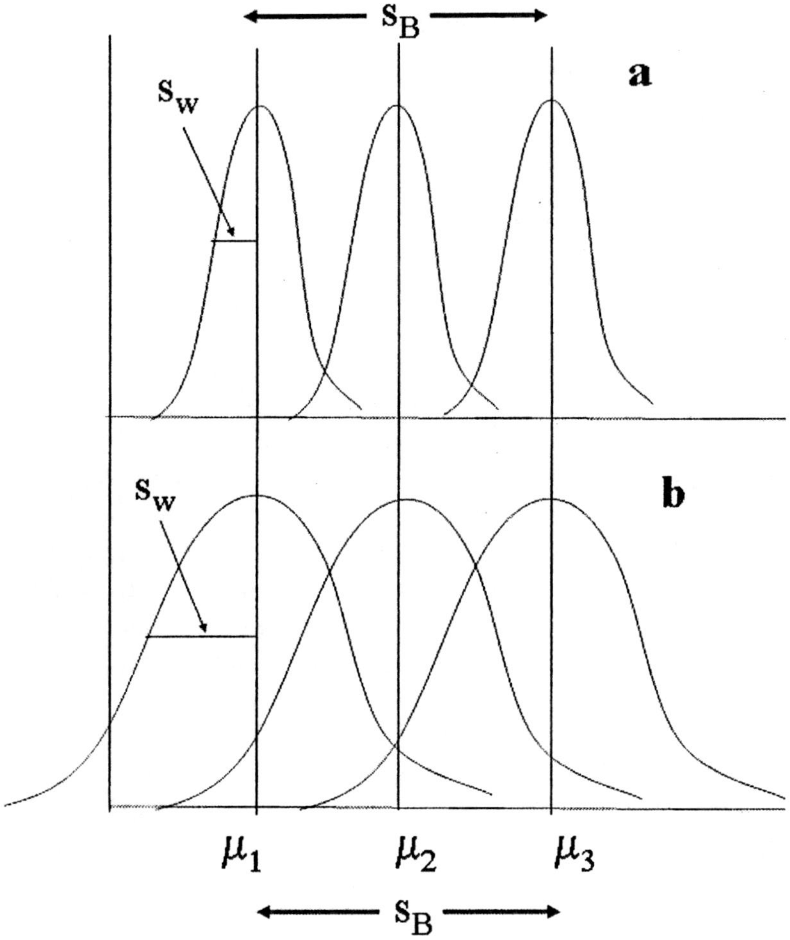

In Figure 8.2 we again show two examples. This time, there are three groups with the same means, but the samples in the example *a* have smaller variances than the samples in example *b*. Even though we now have three samples in each example (instead of two), the within-group variance is the same concept for ANOVA as it was for an independent t-test. However, we can no longer just subtract the two means to determine the difference between them, so we now determine the variance among the means, or the "between group variance, s_B^2". The between-group variance (s_B) is the average squared deviation of each group mean from the grand mean (overall mean of all subjects regardless of group). While the within-group variance (s_w^2) serves as an estimate of the true population variance (σ^2), the between group variance (s_B^2) includes that same natural variation as well as variability between groups that may be due to the effect of the experimental treatment. These are effects which shift the means away from one another.

s_W^2 = within group variance (reflects natural variation in the population, thus it is an estimate of σ^2). Within group variance could also be called "error" (E) because it is variability that we cannot control, but occurs naturally.

s_B^2 = between group variance (reflects natural variation in the population (E) AND the effect of the independent variable, if any).

Thus, each score is the result of a combined influence of the grand mean (μ), any treatment effect attributable to our independent variable (IV), and unexplained error (E).

Linear or additive equation

Score = μ + IV + E

There are multiple potential influences on the outcome for each score, and those influences can be defined with a linear equation. We will discuss this a bit more in Chapter 11. For now, it is important to focus on whether the effect of our independent variable (IV) is enough to demonstrate a statistically significant difference among our groups. To test for that difference, we will calculate an F-obtained value, where F is the symbol for the ANOVA statistic, and it is expressed with the following equation:

$$F = \frac{s_B^2}{s_W^2} = \frac{estimated\sigma^2 \ \&\ treatment}{estimated\sigma^2}$$

The natural variability (estimated σ^2) cancels itself out in the numerator and denominator:

$$F = \frac{s_B^2}{s_W^2} = \frac{estimated\sigma^2 \,\&\, treatment}{estimated\sigma^2}$$

, so what we have left is the effect of our treatment.

Figure 8.3

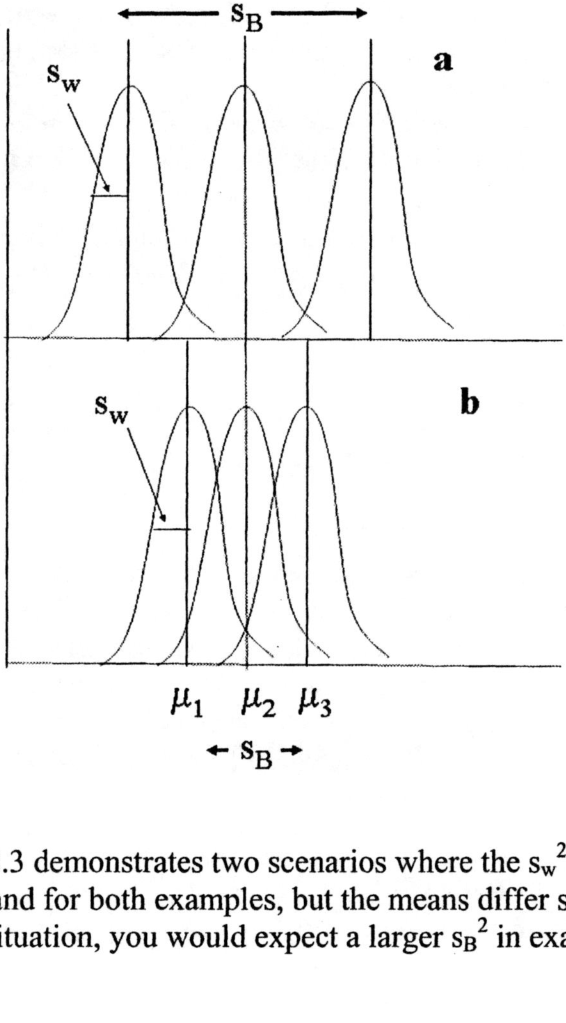

Figure 8.3 demonstrates two scenarios where the s_w^2 is the same for all of the groups and for both examples, but the means differ significantly in example *a*. In this situation, you would expect a larger s_B^2 in example *a* because your

measure of the differences amongst the means will be larger for example *a* than it will be for example *b*. Thus, you are more likely to calculate an F-obtained value that is greater than 1.00 in example *a* than you are for example *b*.

Properties of your F-obtained value

Our F-obtained value should be a measure of our experimental treatment, or the effect of our experiment. If there is no effect, then F-obtained will equal 1.00 (approximately). The reason that F-obtained will equal 1.00 when there is no treatment effect is because you would essentially be dividing your estimated variance by itself, and any number divided by itself is equal to 1.00. Also, F can never have a negative value since it is the ratio of two squared values, and squared values are always positive.

Of course, F-obtained will rarely equal exactly 1.00 since our two estimates of variance are just that, estimates. Each estimate will vary slightly due to random effects, but F-obtained values near 1.00 will never be significant because they will suggest little or no effect of our experimental treatment. However, if we divide the between-groups variance by the within-groups variance and get a number greater than 1.00, then we have a statistic which measures our treatment effect. If the value is large enough to be unlikely to occur by chance, then we can say that our treatment (our independent variable) had a significant effect on our dependent variable. Thus, a larger F-obtained value means a greater effect of the independent variable.

Calculating an F-obtained Value

Recall that variances are calculated dividing the SS by the df. Thus, the between- and within-groups variance calculations involve the appropriate SS and df.

$$s_B^{\,2} = \frac{SS_B}{df_B}$$

$df_B = k - 1$, where k = number of groups/conditions

$$s_W^2 = \frac{SS_W}{df_W}$$

$df_W = N_T - k$, where N_T = total number of subjects in all conditions

$$F_{obtained} = \frac{s_B^2}{s_W^2}$$

Calculating the SS values is done in the usual way by adding and squaring the scores, but it is trickier than it was for the t-tests because there are more than two groups, and you must calculate the ΣX for each group and for all of the subjects in the experiment. Nevertheless, it is conceptually similar to the SS calculations that you have performed in past chapters. You can refer to the sidebars for more information on calculating the SS manually.

We have chosen to give you the SS values to save time, and because you can understand how F-obtained is influenced by variability without manually calculating the SS for each group and for the entire experiment. Keep in mind that the SS_B and the SS_W can be added together to create the SS total (SS_T). Thus, if you know the SS_T and one of the other SS, you can determine the other value by subtraction.

$$SS_T = SS_B + SS_W$$

Table 8.1

ANOVA Summary Table (Sample)

Source	SS	df	s^2 (MS)	F obt.
Between Groups				
Within Groups				
Total				

The formulas listed above can be used to fill out the ANOVA Summary Table. ANOVA results are traditionally presented in this format. By dividing the SS by the appropriate df, you can fill in the variance (or Mean Square) column. Dividing the variance between-groups by the variance within-groups gives you the F-obtained value. Given these "formulas," you can even complete an ANOVA summary table that is only partially complete if you know a few values and the mathematical relationships between them. For example, you could be given the df and variance values, but not the SS values. You could fill in the SS values by multiplying df and variance separately for between- and within-groups. Since $\frac{SS}{df} = s^2$, then $s^2 * df = SS$.

Characteristics of the F distribution

The F-distribution, like the t-distribution, is based on a family of curves that varies with the degrees of freedom. In this case, we have different degrees of freedom for the numerator (df_B) and the denominator (df_W) of the F formula, and there is a different curve for every possible combination of df_B and df_W. As you can imagine, there are unlimited combinations of these two values, and thus we will rely on critical values to indicate the values that F-obtained must equal, or be more extreme than, in order to achieve statistical significance. Because F-obtained always yields a positive number, it is not too surprising that the F distribution is also positive and positively skewed. In addition, all tests of ANOVA are two-tailed. Table C contains the F-critical values that are most likely to be used in this course.

Complete Example

An animal behaviorist measures the length of calls in three closely related species of crows, and asks if there is a difference in call length between the three groups. Assume that the data are normally distributed and the variances are homogeneous. The dependent variable is call length measured in seconds. Let $\alpha = 0.05$.

Song Length (s)

Species A	Species B	Species C
2.5	2.8	3.8
3.1	3.2	3.2
2.2	3.0	4.0
2.6	2.7	3.7
2.2	2.3	3.7

Null Hypothesis: There is no difference in the length of calls between the three groups or the differences are due to chance.

Alternative Hypothesis: There is a difference in the length of the calls of one or more groups from one or more of the other groups.

Step One: Compute the probability that each of the sample means comes from the null hypothesis population where $\mu_1 = \mu_2 = \mu_3$.

Calculate the Means & the Degrees of Freedom for each group

$\Sigma X_A = 12.6$ $\Sigma X_B = 14$ $\Sigma X_C =$ 18.4

$\overline{X}_A = \frac{12.6}{5} = 2.52$ $\overline{X}_B = \frac{14}{5} = 2.80$

$\overline{X}_C = \frac{18.4}{5} = 3.68$

$df_B = k - 1$, where k = number of groups/conditions, so $df_B = 3 - 1 = 2$
$df_W = N_T - k$, where N_T = total number of subjects in all conditions, so $df_W = 15 - 3 = 12$

*Note: You can check your work on the degrees of freedom by calculating the df_T.
$df_T = N_T - 1 = df_B + df_W$, so $df_T = 15 - 1 = 2 + 12 = 14$

Apply the s^2 & F-obtained formulas to complete the ANOVA Summary Table

Table 8.2

Crow Call Length Example
ANOVA Summary Table

Source	SS	df	s^2 (MS)	F obt.
Between Groups	3.664	2	**1.832**	**16.212**
Within Groups	1.356	12	**0.113**	
Total	**5.020**	14		

Numbers in bold were filled in by solving the formulas while df (not bolded) were calculated in the previous step. Calculations for the SS are available in the sidebar.

$SS_T = SS_B + SS_W$

$SS_T = 3.664 + 1.356 = 5.020$

$$s_B^2 = \frac{SS_B}{df_B} = \frac{3.664}{2} = 1.832$$

$$s_W^2 = \frac{SS_W}{df_W} = \frac{1.356}{12} = 0.113$$

$$F_{\text{obtained}} = \frac{s_B^2}{s_W^2} = \frac{1.832}{0.113} = 16.212$$

Step Two: Evaluating the probability of obtaining this score due to chance

Evaluate the F-obtained value based on alpha (α) = 0.05. To evaluate your F-obtained value, you must use the F distribution (Table C), and to do so, you need an alpha level (0.05) and degrees of freedom (df_B and df_W). The degrees of freedom from the numerator of the F-obtained equation are the df_B while the degrees of freedom from the denominator are df_W. Thus, the df for this problem are 2 & 12. Compare the F-critical value with your F-obtained value, and reject the null hypothesis when your F-obtained value is equal to or greater than your F-critical value. When $\alpha = 0.05$ and the degrees of freedom are equal to 2 & 12, you should use 3.88 as your F-critical value. $16.212 \geq 3.88$, so we *reject* the null hypothesis.

How should we interpret these data in light of the call lengths in three species of crows?

These results suggest that you would expect to obtain these

differences in call length less than 5% of the time, if species membership had no effect. Thus, it is not very likely that these call length differences all come from the normal null hypothesis population. Of course, as always, there is some finite probability that you could get a difference this large among three means that is purely due to chance, but that probability is less than 5%.

In addition, we can conclude that one or more of these three means is significantly different from one or more of the other means. We also know from our calculations of the means above that species C has the longest call length and that species A has the shortest call length, so it is likely that we have a significant difference between these two groups, but what about B vs. C, or A vs. B?

Determining the statistical significance of these pairwise comparisons can be done using "multiple comparisons" once you know that your overall test (your ANOVA) has demonstrated statistical significance. We will turn to this concept in the next section of this chapter, but we will first take a look at the Excel and SPSS output for these data that we have analyzed by hand.

Results if you use Microsoft Excel to calculate the ANOVA

Anova: Single Factor

SUMMARY

Groups	*Count*	*Sum*	*Average*	*Variance*
Species A	**5**	**12.6**	**2.52**	0.137
Species B	**5**	**14**	**2.8**	0.115
Species C	**5**	**18.4**	**3.68**	0.087

ANOVA

Source of Variation	*SS*	*df*	*MS*	*F*	*P-value*	*F crit*
Between Groups	3.664	**2**	**1.832**	**16.21239**	0.000388	**3.88529**
Within Groups	1.356	**12**	**0.113**			
Total	**5.02**	**14**				

This is the output that you get from Excel when you type in the data for the three species of crows. Note that Excel calls this test an "ANOVA: Single Factor". This wording is slightly different than we have been using, but it is describing exactly the same analysis; this will be even clearer after we have discussed other types of ANOVA analyses in the next chapter. We have bolded the numbers that are comparable to the numbers that we just manually calculated or looked up in the table. Note that the F-obtained value (Excel calls it "F") is identical to ours, except they have carried more digits (16.21239). The "F Crit" is also the same (3.88529), but again they have carried more digits. The degrees of freedom are the same (2 & 12). In addition, they have provided you with the exact probability ("P-value") of obtaining an F-obtained value ("F") of 16.21239 which is 0.000388. Since 0.000388 < 0.05, we clearly *reject* our null hypothesis. These results suggest that exactly 0.0388% of the time, you would obtain this difference in call lengths if species membership had no effect. Again, knowing the exact probability (not just the F-critical values) for each F-obtained value is one of the major advantages of using a computer to calculate your analyses.

Results if you use SPSS to calculate the ANOVA (or F test)

ANOVA

CALLS

	Sum of Squares	df	Mean Square	F	Sig.
Between Groups	3.664	**2**	**1.832**	**16.212**	.000
Within Groups	1.356	**12**	**.113**		
Total	**5.020**	**14**			

We have placed the numbers that are comparable to our manual calculations in bold. Once again, you see that the exact probability (SPSS calls it "Sig.") is less than our alpha level of 0.05, thus we must assume that these results would be *unlikely* simply due to chance.

How would these results be reported in a scientific journal article?

If the journal required American Psychological Association (APA) format, the results would be reported in a format something like this:

> There is a significant difference in the length of the calls of one or more groups from one or more of the other groups, $F(2,12) = 16.21, p < .01$.

This formal sentence includes the dependent variable (call length), the independent variable (three groups or species), as well as a statement about statistical significance, the symbol of the test (F), the degrees of freedom between groups (2), the degrees of freedom within groups (12), the statistical value (16.21), and the probability of obtaining this result simply due to chance (<.01).

Another Complete Example

Teachers in the K-12 grades all too frequently "burn out" and leave the teaching profession. A psychologist suggests that a combination (one or more) of four assessments of stress and professional efficacy might provide the option to anticipate such problems and take action to prevent losing good teachers. The immediate question is whether scores from the four assessments are all the same across teachers, or whether there is a difference in scores for the assessments. A group of 40 teachers are randomly assigned to four equal groups and each group is given one of the assessments. Is there a significant difference in mean scores for the four groups? Assume that the data are normally distributed and the variances are homogeneous. The dependent variable is assessment score. Let $\alpha = 0.05$.

Assessment Scores

Group A	Group B	Group C	Group D
16.3	18.3	13.6	16.7
23.2	21.3	14.6	18.2
21.0	19.5	19.0	18.6
19.0	18.6	21.0	19.5
21.9	21.4	14.9	12.6
16.4	18.9	21.9	13.7
14.9	12.6	21.9	21.4
21.9	13.7	16.4	18.9
13.6	16.7	16.3	18.3
14.6	18.2	23.2	21.3

Null Hypothesis: There is no difference in the assessment scores among the four groups or types of assessments, or the differences are due to chance.

Alternative Hypothesis: There is a difference in the mean assessment scores obtained from one or more of the assessments.

Step One: Compute the probability that each of the sample means comes from the null hypothesis population where $\mu_1 = \mu_2 = \mu_3 = \mu_4$.

Calculate the Means & the Degrees of Freedom for each group

$\Sigma X_A = 182.8$ $\Sigma X_B = 179.2$ $\Sigma X_C = 176.7$ $\Sigma X_D = 185.3$

$\bar{X}_A = 18.28$ $\bar{X}_B = 17.92$ $\bar{X}_C \doteq 17.67$ $\bar{X}_D = 18.53$

$df_B = k - 1$, where k = number of groups/conditions, so $df_B = 4 - 1 = 3$

$df_W = N_T - k$, where N_T = total number of subjects in all conditions, so $df_W = 40 - 4 = 36$

*Note: You can check your work on the degrees of freedom by calculating the df_T.

$df_T = N_T - 1 = df_B + df_W$, so $df_T = 40 - 1 = 3 + 36 = 39$

Apply the s^2 & F-obtained formulas to complete the ANOVA Summary Table

Table 8.3

Burnout Assessment Example
ANOVA Summary Table

Source	SS	df	s^2 (MS)	F obt.
Between Groups	4.35	3	**1.45**	**0.14**
Within Groups	372.81	36	**10.36**	
Total	**377.16**	39		

Numbers in bold were filled in by solving the formulas while df (not bolded) were calculated in the previous step.

$SS_T = SS_B + SS_W$

$SS_T = 4.35 + 372.81 = 377.16$

$$s_B^2 = \frac{SS_B}{df_B} = \frac{4.35}{3} = 1.45$$

$$s_W^2 = \frac{SS_W}{df_W} = \frac{372.81}{36} = 10.36$$

$$F_{obtained} = \frac{s_B^2}{s_W^2} = \frac{1.45}{10.36} = 0.14$$

Step Two: Evaluating the probability of obtaining this score due to chance

Evaluate the F-obtained value based on alpha (α) = 0.05. To evaluate your F-obtained value, you must use the F distribution (Table C), and to do so, you need an alpha level (0.05) and degrees of freedom (df_B

and df_W). The degrees of freedom from the numerator of the F-obtained equation are the df_B while the degrees of freedom from the denominator are df_W. Thus, the df for this problem are 3 & 36. Compare the F-critical value with your F-obtained value, and reject the null hypothesis when your F-obtained value is equal to or greater than your F-critical value. When $\alpha = 0.05$ and the degrees of freedom are equal to 3 & 36, you should use 2.866 as your F-critical value. $0.14 < 2.86$, so we *fail-to-rejectt* the null hypothesis.

How should we interpret these data in light of the differences in assessment outcomes?

These results suggest that you would expect to obtain these differences in assessment outcomes more than 5% of the time, if there was no difference in the assessments. Thus, it is very likely that these assessment outcomes all come from the same normal null hypothesis population. Of course, as always, there is some finite probability that there is a difference in the outcomes of one or more of these assessments that you are not detecting with this analysis.

Unlike the previous example, there is no need to determine the statistical significance of the pairwise comparisons among these four means, since the overall F-test has failed to reject the null hypothesis. If the overall test does not detect a difference among your means, you DO NOT continue to the pairwise comparisons of the means: you are done with the analysis. Let's take a look at the Excel and SPSS output for these data that we have analyzed by hand.

Results if you use Microsoft Excel to calculate the ANOVA

SUMMARY

Groups	*Count*	*Sum*	*Average*	*Variance*
Group A	**10**	**182.8**	**18.28**	12.517
Group B	**10**	**179.2**	**17.92**	8.364
Group C	**10**	**176.7**	**17.67**	10.142
Group D	**10**	**185.3**	**18.53**	10.400

ANOVA

Source of Variation	*SS*	*df*	*MS*	*F*	*P-value*	*F crit*
Between Groups	4.346	**3**	**1.449**	**0.140**	0.935	**2.866**
Within Groups	372.814	**36**	**10.356**			
Total	**377.160**	**39**				

This is the output that you get from Excel when you type in the data for the four assessment groups. We have bolded the numbers that are comparable to the numbers that we just manually calculated or looked up in the table. Note that these results are the same as our hand-calculated example. In addition, they have provided you with the exact probability ("P-value") of obtaining an F-obtained value ("F") of 0.14, which is 0.935. Since 0.935 > 0.05, we clearly *fail-to-reject* our null hypothesis. These results suggest that exactly 93.5% of the time, you would obtain this difference in assessment scores if the assessments were not different.

Results if you use SPSS to calculate the ANOVA (or F test)

ANOVA
SCORES

	Sum of Squares	df	Mean Square	F	Sig.
Between Groups	4.346	**3**	**1.449**	**.140**	.935
Within Groups	372.814	**36**	**10.356**		
Total	**377.160**	**39**			

We have placed the numbers that are comparable to our manual calculations in bold. Once again, you see that the exact probability (SPSS calls it "Sig.") is greater than our alpha level of 0.05, thus we must assume that these results would be *likely* due to chance, and therefore do not reflect real differences among the assessments.

How would these results be reported in a scientific journal article?

If the journal required American Psychological Association (APA) format, the results would be reported in a format something like this:

> There is no significant difference in the assessment scores among the four groups or types of assessments, $F(3,36) = .14$, $p = .93$.

This formal sentence includes the dependent variable (assessment scores), the independent variable (four groups), as well as a statement about statistical significance, the symbol of the test (F), the degrees of freedom between groups (3), the degrees of freedom within groups (36), the statistical value (.14), and the probability of obtaining this result simply due to chance (.93).

When is it appropriate to use an ANOVA test?

1) Use an independent ANOVA when you have more than two samples and a between groups design, and
2) When the sampling distribution is normally distributed. Again, this is satisfied when:
   a. The sample size is greater or equal to 30 in each group ($n \geq 30$) or,
   b. The null hypothesis population is known to be normally distributed.
   and
3) When the dependent variable is on an interval or ratio scale (although some statisticians believe that this is too strict), and
4) When the variances of the groups are the same, or are **homogeneous**. The homogeneity of variance assumption (HOV) requires that the variances of the samples are equal, or in practical terms, not significantly different from one another. We will use a simple rule of thumb to decide whether the variances are similar enough to be considered homogeneous:

Rule of Thumb for HOV check: if the $\frac{\text{LargestVariance}}{\text{SmallestVariance}} \leq 4$ then HOV assumption is met (okay) and you can proceed with ANOVA. If not, you cannot proceed with the ANOVA (later, we'll explain what to do instead).

Multiple Comparisons

If you are fortunate enough to obtain statistical significance with your ANOVA (F test), then you will need to do multiple comparisons to determine which of your means is significantly different from which of your other means. There are two major categories of comparisons, planned and unplanned. We will overview how to do the planned comparisons first, and then move to the more-common unplanned or post-hoc comparisons.

Planned (a priori) Comparisons

Planned comparisons are performed whenever you have pre-planned a *limited* number of comparisons prior to collecting your data. Planned comparisons are also called "a priori" comparisons because "a priori" is a Latin term that means "from what goes before; from cause to effect." Because you have pre-planned these comparisons, typically based on prior data and theory, and you do not plan to do *all possible* comparisons, you are not required to make a correction for your alpha (α) level. In this situation, it is appropriate to conduct independent t-tests to determine if your planned comparisons have means that are significantly different from one another. Fortunately, this calculation is simplified based on the calculations you have already done for the ANOVA summary table. You are able to obtain the s_w^2 (a better estimate of variance than just taking the average of two estimates) directly from the ANOVA summary table to insert into your independent t-test formula:

$$t_{obtained} = \frac{\overline{X}_1 - \overline{X}_2}{\sqrt{s_w^{\,2}(1/n_1 + 1/n_2)}}$$

From the ANOVA Summary Table

Crow Example

If we revisit the crow call analysis we performed as an ANOVA, we can now determine whether or not species A vs. C are significantly different and whether or not species B vs. C are significantly different from one another by performing two independent t-tests. Note that we are *not* performing a t-test on all possible comparisons because we are not comparing A vs. B. We would have had to pre-plan the two comparisons that we are performing if this had been a real study. For your convenience we have repeated the ANOVA summary table from SPSS to refer to as we demonstrate the calculations.

$$t_{obtained} = \frac{\overline{X}_1 - \overline{X}_2}{\sqrt{s_w^{\,2}(1/n_1 + 1/n_2)}}$$

Anova: Single Factor

SUMMARY

Groups	*Count*	*Sum*	*Average*	*Variance*
Species A	5	12.6	2.52	0.137
Species B	5	14	2.8	0.115
Species C	5	18.4	3.68	0.087

ANOVA

Source of Variation	*SS*	*df*	*MS*	*F*	*P-value*	*F crit*
Between Groups	3.664	2	1.832	16.21239	0.000388	3.88529
Within Groups	1.356	12	0.113			
Total	5.02	14				

Step One: Calculate the t obtained values for Species A vs. C and B vs. C

$$t_{obtained} = \frac{2.52 - 3.68}{\sqrt{0.113(1/5 + 1/5)}} = \frac{-1.16}{\sqrt{0.113(.40)}} = \frac{-1.116}{0.2126} = \text{-5.46 (A vs. C)}$$

$$t_{obtained} = \frac{2.80 - 3.68}{\sqrt{0.113(1/5 + 1/5)}} = \frac{-0.88}{\sqrt{0.113(.40)}} = \frac{-0.88}{0.2126} = \text{-4.14 (B vs. C)}$$

Step Two: Evaluating the probability of obtaining this score due to chance

Evaluate the t-obtained value based on alpha (α) = 0.05, and a two-tailed hypothesis. To evaluate your t-obtained value, you must use the t distribution (Table B). To determine your t-critical value, you need to know your alpha level (0.05), the number of tails you are evaluating (2 in this case) and your degrees of freedom (df). We use the degrees of freedom associated with the s_w^2, or the df_w, to evaluate the t-obtained value. Thus, the df for this problem are 12. Compare the t-critical value with your t-obtained value. When $\alpha = 0.05$, your degrees of freedom are equal to 12, and your hypothesis is two tailed, you should use 2.179 as your t-critical value.

$-5.46 \geq |2.179|$, so we *reject* the null hypothesis that there is no difference between species A and C.

$-4.14 \geq |2.179|$, so we *reject* the null hypothesis that there is no difference between species B and C.

How should we interpret these data in light of the effect of species on the length of calls?

These results suggest that less than 5% of the time, you would obtain these differences in call length if there was no effect of species membership. Thus, it is not very likely that these call lengths for species A vs. C and B vs. C come from the normal null hypothesis population of mean differences. However, there is a chance that you could get a difference this large between two means that is purely due to chance, but that chance is less than 5%.

How would these results be reported in a scientific journal article?

If the journal required American Psychological Association (APA) format, the results would be reported in a format something like this:

> There is a significant difference in the call lengths of species A and C, $t(12) = -5.46, p < .05$.
>
> There is a significant difference in the call lengths of species B and C, $t(12) = -4.14$, $p < .05$.

These formal sentences includes the dependent variable (call length), the independent variable (species), as well as a statement about statistical significance, the symbol of the test (t), the degrees of freedom within groups (12), the statistical value (-.546 and -4.14), and the probability of obtaining these results simply due to chance (<.05).

Unplanned (a posteriori) or Post-hoc Comparisons

Unplanned comparisons are often called post-hoc or "a posteriori" comparisons. "A posteriori" is a Latin phrase that means "from what follows; from effect to cause." "Post hoc" is an abbreviation for a Latin phrase "*post hoc ergo propter hoc*" which means "after this therefore because of this". This phrase describes the fallacy that can occur when you assume that because one thing follows another that the one thing was caused by the other. Of course, not planning a limited number of comparisons prior to the start of the study also means that you must perform comparisons that limit your type I error rate. While this is less desirable than planned comparisons, it is often necessary when there is no strong prior knowledge regarding which groups are likely to differ.

There are many different techniques that limit your type I error rate. Some of these techniques are considered liberal (meaning that it easier to reject the null hypothesis), while others are considered conservative (harder to reject the null hypothesis). Because there are many different post-hoc tests, and most computer statistical packages will do a variety of them, we will focus on one to calculate manually to demonstrate how the test can limit your probability of making a type I error. The test that we will show is the Newman-Keuls test. This test maintains your alpha level by adjusting the critical value for the number of means being compared (r). The number of means being compared is determined by ordering all of the means from smallest to largest, and then determining for each comparison whether or not there are means that fall between the two target means. For example, what if you had four means and you wanted to make every possible comparison between them? If you had four means then you would need to perform six post hoc comparisons, which would be:

1 vs. 2 AND 1 vs. 3 AND 1 vs. 4 AND 2 vs. 3 AND 2 vs. 4 AND 3 vs. 4.

Before you could determine how many means were encompassed (included) in each comparison, you would need to rank your means from smallest to largest.

From:

$\overline{X_1} = 33.1$ $\overline{X_2} = 18.8$ $\overline{X_3} = 36.2$ $\overline{X_4} = 4.9$

To:

$\overline{X_4} = 4.9$ $\overline{X_2} = 18.8$ $\overline{X_1} = 33.1$ $\overline{X_3} = 36.2$

So, when you make the comparison between the means of group 4 vs. 3, you are actually encompassing two means in the middle (r = 4 when the means are ranked in order of their values). But, when you compare groups 1 vs. 3 there are no means in between them, and r = 2.

Q Distribution

The Newman-Keuls test, like the statistical tests you have already learned, has its own sampling distribution that is used to evaluate whether our obtained values are likely due to chance alone. The Q distribution was created by randomly taking samples of equal n and calculating the difference between the most extreme means. This created a distribution that tells us how often you can get two sample means that encompass "r" means from the same population and have them differ by the amount calculated with your Newman-Keuls test. As we have done before, we will evaluate our results so that we reject the null hypothesis suggesting that there is no difference between the means if the Q statistic is less than a critical value of Q..

Formula for Calculating a Newman-Keuls Comparison (Q-obtained)

$$\text{Q-obtained} = \frac{\overline{X_i} - \overline{X_j}}{\sqrt{\frac{s_w^2}{n}}}$$

Where $\overline{X_i}$ = Larger of the Two Means

Where $\overline{X_j}$ = Smaller of the Two Means

Note: If your n's are not equal, you can calculate a harmonic mean of your sample sizes and use the harmonic mean in the formula. Because this is not critical to your understanding of post hoc concepts, we limit our examples and exercises to situations where the sample sizes are equal.

① **Complete Example**

Let's return to our crow call example and analyze all possible comparisons of the three means using the Newman-Keuls approach. Evaluate the results based on $\alpha = 0.05$.

Step One: Rank order the means

$\overline{X}_A = 2.52$ $\overline{X}_B = 2.80$ $\overline{X}_C = 3.68$

Simply by chance, the means are already in the correct order!

Step Two: Calculate Q-obtained values for all possible comparisons

$$\text{Q-obtained} = \frac{\overline{X}_i - \overline{X}_j}{\sqrt{\frac{s_w^2}{n}}}$$

$$\text{A vs. B: Q-obtained} = \frac{2.80 - 2.52}{\sqrt{\frac{0.113}{5}}} = 1.863 \qquad r = 2$$

$$\text{A vs. C: Q-obtained} = \frac{3.68 - 2.52}{\sqrt{\frac{0.113}{5}}} = 7.718 \qquad r = 3$$

$$\text{B vs. C: Q-obtained} = \frac{3.68 - 2.8}{\sqrt{\frac{0.113}{5}}} = 5.855 \qquad r = 2$$

Step Three: Evaluate Q-obtained based on Q-critical values

Evaluate the Q-obtained values based on alpha (α) = 0.05, the number of means that your comparisons encompass (in this example either two or three means), and the df_w. To evaluate your Q-obtained values, you must use the Q distribution (Table D). When $\alpha = 0.05$, your degrees of freedom are equal to 12, and $r = 2$, you should use 3.08 as your Q-critical value. When $r = 3$, you should use 3.77 as your critical value.

$1.863 < 3.08$, so we *fail-to-reject* the null hypothesis that there is no difference between species A and B.

$7.718 \geq 3.77$, so we *reject* the null hypothesis that there is no difference between species A and C.

$5.855 \geq 3.08$, so we *reject* the null hypothesis that there is no difference between species B and C.

How should we interpret these data in light of the effect of species on call length?

These results suggest that for Species A vs. C and B vs. C, there is less than a 5% chance that you would obtain these differences in call length if there was no effect of species membership. Thus, it is not very likely that these call lengths for species A vs. C and B vs. C come from the normal null hypothesis population of mean differences. However, there is a chance that you could get a difference this large between two means that is purely due to chance, but that chance is less than 5%.

Results if you use SPSS to calculate the Newman-Keuls

CALLS
Student-Newman-Keuls

	N	Subset for alpha = .05	
SPECIES		1	2
1.00	5	**2.5200**	
2.00	5	**2.8000**	
3.00	5		**3.6800**
Sig.		.212	1.000

Means for groups in homogeneous subsets are displayed.
a Uses Harmonic Mean Sample Size = 5.000.

This is the output you get from SPSS when you type in the data for the three species of crows. Note that SPSS calls this test a "Student-Newman-Keuls or SNK". The numbers in bold are the means of the three samples, and SPSS has divided the means into subgroups which cannot be distinguished statistically at the stated alpha level (0.05 in this example). So, the means for samples 1 and 2 are not significantly different, but the mean for sample 3 is significantly different from the means for samples 1 and 2. This is the same conclusion that we obtained in doing the problem by hand.

How would these results be reported in a scientific journal article?

If the journal required American Psychological Association

(APA) format, the results would be reported in a format something like this:

> There is no significant difference in the call lengths of species A and B, $Q(12) = 1.86, p > .05$.
>
> There is a significant difference in the call lengths of species A and C, $Q(12) = 7.72, p < .05$.
>
> There is a significant difference in the call lengths of species B and C, $Q(12) = 5.86$, p $< .05$.

These formal sentences includes the dependent variable (call length), the independent variable (species), as well as a statement about statistical significance, the symbol of the test (Q), the degrees of freedom within groups (12), the statistical value (1.86, 7.72 and 5.86), and the probability of obtaining these results simply due to chance (>.05, <.05, <.05).

Overview of Single Sample, Two Sample & 3+ Sample Tests: Expanded Flowchart

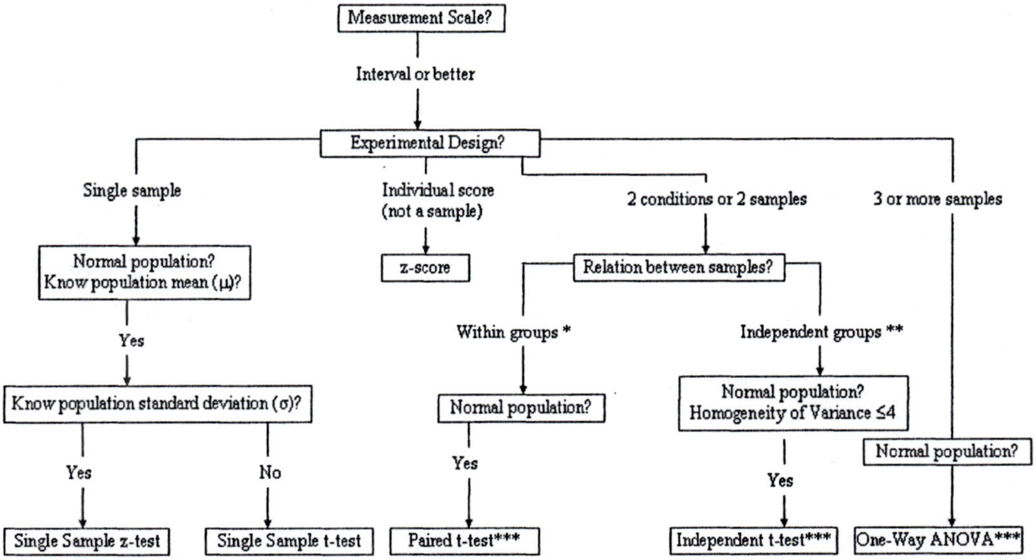

1. *Within Groups
   a. subjects matched - married, partners, twins, etc., or individually matched for a reason that might affect the I.V.
   b. before vs. after conditions with the same subjects, measured both times
2. **Independent Groups
   a. In one group/condition, NOT allowed in the other, subjects randomly assigned to be in the experiment in one group OR the other
   b. Sample sizes can be unequal (IF unequal, definitely independent t-test, but if equal, have to make decision based on something else).

Notice that the flowchart does not yet tell you what to do if you do not meet the assumptions of your statistical tests. In fact, there are some options when you have three samples, but you do not meet the assumptions for parametric tests. The one-way ANOVA can be replaced with a non-parametric test called the Kruskal-Wallis test. Another option is to perform a log transformation on your data to determine if the mathematical transformation will result in a normal distribution.

Summary

Chapter 8: Analysis of Variance

In this chapter you have covered statistical tests for designs in which you have three or more samples in an independent, or between-groups, design, and you lack information about the underlying population. We have shown you the steps involved in conducting your overall analyses (F test) as well as the steps involved in determining where your means are significantly different (providing that your first overall F test is significant). In the next chapter, we will extend the concept of an independent one-way ANOVA design to more complex situations where we have a within-groups design or more than one factor (independent variable).

Chapter Eight Homework

Provide a short answer for the following questions.

1. How does variability of the samples, in all of its forms, influence the F-obtained value?
2. The F distribution has no negative values. True or false? How do you account for this?
3. Why do we have to use a Newman-Keuls test instead of multiple independent t-tests?
4. If $F<1$, we know that there was no significant effect of our independent variable. How do we know this?
5. How is the F distribution similar to the t distribution?
6. What are the assumptions for one-way ANOVA?
7. How do you check homogeneity of variance when conducting an ANOVA test?

8. A behavioral neuroscientist tests the effects of three drugs on motor coordination and obtains motor coordination scores for each subject. There are three groups with four subjects in each condition. Assume that the population of scores is normally distributed and all variances are homogeneous. Complete the ANOVA source table provided below. Using an $\alpha = 0.05$, what do you conclude? Reject or Fail-to-Reject the H_0 and perform a Newman-Keuls test, if appropriate.

Drug A: $\overline{X} = 9$
Drug B: $\overline{X} = 9.5$
Drug C: $\overline{X} = 15.25$

Source	SumsSquares	df	s^2	F_{obt}	F_{crit}
Between	148.17	___	_____	________	________
Within	__________	___	7.1944		
Total	__________	___			

9. A research team specializing in Borderline Personality Disorder tests three new drugs to treat the condition. Eighty-four volunteers with Borderline Personality Disorder are divided equally and randomly into 4 groups. Groups A – C each receive different experimental drugs to treat the disorder along with behavioral therapy and Group D receives a placebo and behavioral therapy. Levels of a neurotransmitter known to be associated with the disorder are assessed after six months of therapy and the groups are compared. Assume that the population of scores are normally distributed and all variances are homogeneous. Test the hypothesis that the drug affects neurotransmitter levels. Complete the following analysis of variance table, reject or fail-to reject the H_0, and conduct a Newman-Keuls test if appropriate:

Group A: $\overline{X} = 155.25$
Group B: $\overline{X} = 123.33$
Group C: $\overline{X} = 59.25$
Group D: $\overline{X} = 165.99$

Source	SumsSquares	df	s^2	F_{obt}	F_{crit}
Between	3.2661	____	____	______	______
Within	________	____	____		
Total	40.5061	____			

CHOOSE THE MOST APPROPRIATE AND POWERFUL TEST:

10. Several studies indicate that handedness (left-handed/right-handed) is related to differences in brain function. Because different parts of the brain are specialized for specific behaviors, this means that left-handed and right-handed people should show different skills or talents. To test this hypothesis, a psychologist tested pitch discrimination for three

randomly selected groups of subjects: 34 left-handed, 31 right-handed and 30 ambidextrous. Pitch discrimination is assessed based on the subject's ability to determine if the chord is a third (notes separated by four half steps) or a fifth (notes separated by seven half steps). Variances are 1.8, 2.4, and 3.2 respectively.

11. Research on human memory indicates that, on average, people can remember 5 pieces of new information at a time. A random sample of 16 subjects from a normally distributed population trained in mnemonics were read a list of 20 words to remember, then asked to count backwards from 100 for a minute, then asked to recall as many of the words as possible. Their mean recall was 8 words with a standard deviation of 2 words. Using this information, determine whether the subjects in this sample were representative of the general population in which the average recall is known to be 5 words.

12. Over a 20 year period, first-year students at a state university had a mean grade-point average of 2.27 with a standard deviation of 0.42. Recently, the university has tried to increase admissions standards. A sample of 30 first-year students accepted into the university this year had a mean GPA of 2.63. Have the new admissions standards resulted in a significant change in GPA of first year students?

13. A researcher is interested to know whether there is any difference in mathematical aptitude for six year old boys and six year old girls. Thirty-five children of both sexes were randomly selected to participate in the study. Both groups were given a math skills test in which the number of questions answered correctly were evaluated. Girls averaged a mean of 23 problems right with a standard deviation of 4. Boys averaged a mean of 27 problems right with a standard deviation of 6. Can the researcher conclude that six year old boys and girls differ in their mathematical aptitude?

Sidebar formulas:

$$SS_W = \sum^{\substack{all \\ scores}} X^2 - \left[\frac{(\Sigma X_1)^2}{n_1} + \frac{(\Sigma X_2)^2}{n_2} + \frac{(\Sigma X_3)^2}{n_3} + \ldots + \frac{(\Sigma X_k)^2}{n_k}\right]$$

$$SS_B = \left[\frac{(\Sigma X_1)^2}{n_1} + \frac{(\Sigma X_2)^2}{n_2} + \frac{(\Sigma X_3)^2}{n_3} + \ldots + \frac{(\Sigma X_k)^2}{n_k}\right] - \frac{\left(\sum^{\substack{all \\ scores}} X\right)^2}{N}$$

Where $\sum^{\substack{all \\ scores}} X^2$ = sum of all of the squared scores (regardless of group/condition)

$\sum^{\substack{all \\ scores}} X$ = sum of all of the scores (regardless of group/condition)

ΣX_1 = sum of the scores for group 1
n_1 = number of scores in group 1
ΣX_2 = sum of the scores for group 2
n_2 = number of scores in group 2
ΣX_3 = sum of the scores for group 3
n_3 = number of scores in group 3
ΣX_k = sum of the scores for group k (last group)
n_k = number of scores in group k
N = total number of scores (regardless of group/condition) or sum of $n_1+n_2+n_3+\ldots+n_K$

Sidebar SS for species

$$\sum^{\substack{all\\scores}} X^2 = 140.02$$

$$\sum^{\substack{all\\scores}} X = 45$$

$\Sigma X_1 = 12.6$

$\Sigma X_2 = 14$

$\Sigma X_3 = 18.4$

$n_1 = 5$

$n_2 = 5$

$n_3 = 5$

$N = 15$

$$SS_W = \sum^{\substack{all\\scores}} X^2 - \left[\frac{(\Sigma X_1)^2}{n_1} + \frac{(\Sigma X_2)^2}{n_2} + \frac{(\Sigma X_3)^2}{n_3} + \ldots + \frac{(\Sigma X_k)^2}{n_k}\right]$$

$$SS_W = 140.02 - \left[\frac{(12.6)^2}{5} + \frac{(14)^2}{5} + \frac{(18.4)^2}{5}\right]$$

$$SS_W = 140.02 - \left[\frac{158.76}{5} + \frac{196}{5} + \frac{338.56}{5}\right]$$

$$SS_W = 140.02 - [31.752 + 39.2 + 67.712]$$

$$SS_W = 140.02 - 138.664 = 1.356$$

$$SS_B = \left[\frac{(\Sigma X_1)^2}{n_1} + \frac{(\Sigma X_2)^2}{n_2} + \frac{(\Sigma X_3)^2}{n_3} + \ldots + \frac{(\Sigma X_k)^2}{n_k}\right] - \frac{\left(\sum^{\substack{all\\scores}} X\right)^2}{N}$$

$$SS_B = \left[\frac{(12.6)^2}{5} + \frac{(14)^2}{5} + \frac{(18.4)^2}{5}\right] - \frac{(45)^2}{15}$$

$$SS_B = [31.752 + 39.2 + 67.712] - \frac{2025}{15}$$

$$SS_B = 138.664 - 135 = 3.664$$

Chapter 9
Complex ANOVA Designs

"Complex ANOVA designs" may sound intimidating, but it's really just an extension of the last chapter. Some of the modifications involve adding another independent variable (factor) to the design while others are adjustments that you make when you analyze two or more "paired" or repeated or within-groups (same subjects at different times, for example).

In our discussion of ANOVA in the last chapter, we suggested that each score is the result of a combined influence of the grand mean (μ), any treatment effect attributable to our independent variable (IV), and unexplained error (E). This was represented by a linear or additive equation:

$$\text{Score} = \mu + \text{IV} + \text{E}$$

In this chapter we will modify that conceptual equation to account for additional independent variables (or factors) and for situations in which you use a within-groups experimental design.

Two Factor Designs and Interactions

What do you do if you have more than one independent variable in your experiment that you need to analyze? Your first thought might be that you could perform two one-way ANOVA's, or one ANOVA for each independent variable. The problem with this approach is the same one that we faced when we discussed doing multiple independent t-tests when you have three or more groups: the issue of Type I error. The more tests you perform, the greater your probability of making a Type I error (rejecting the null hypothesis when it is, in fact, true). The solution to the problem is to simultaneously analyze your independent variables in one overall analysis (F test). To do that, we need a single test for two independent variables, and under the correct conditions, the test that fits those requirements is a Two-way ANOVA. One of the additional advantages of using this test is that you can evaluate whether there is an interaction between your two independent variables. This concept of interactive effects is a new one.

Applying the Two-Way ANOVA Concept to our Crow Example

If we continue to use the crow call example with which you are familiar from the last chapter, then making that design a two-factor design could be done as easily as adding information on sex for each subject. Thus, we would be asking, "What are the effects of sex and of species on the call length of crows?" One factor would be sex, and all of the data would be averaged based on sex alone (male or female), while the other factor would be species, and the influence of species (A or B or C) would be calculated while ignoring sex. As a result of this analysis, you can get an idea of the relative influence of one factor relative to another factor. These influences are technically termed "main effects", so that you have a "main effect" for species and a "main effect" for sex. Sometimes the influences of the two independent variables are completely independent of one another, but sometimes they are not. When the effect of one of your independent variables (or factors) is not the same at all levels of your other independent variable (or factor), then you have a statistical "interaction" effect. Examples of main and interaction effects are illustrated in Figure 9.1. A significant interaction occurs when there is an effect of the combined influence of your independent variables that is over and above their individual independent effects. In Fig 9.1a and 9.1b, the effects of sex are the same for each species (in 9.1a, males have longer calls across all species, and in 9.1b, there is no difference between sexes across all species). However, in Figure 9.1c, there is no sex difference in Species A, but a significant difference between sexes in Species C (sex differences in Species B might, or might not, be signficant). This would represent a significant interaction effect. The significant interaction effect is similar in Figure 9.1d, except that there is clearly no sex effect in Species A or Species B. The significant interaction effect in Figure 9.1e is most dramatic, since the effect of sex is in the opposite direction in Species A and Species C, and there is no sex difference in Species B.

Figure 9.1

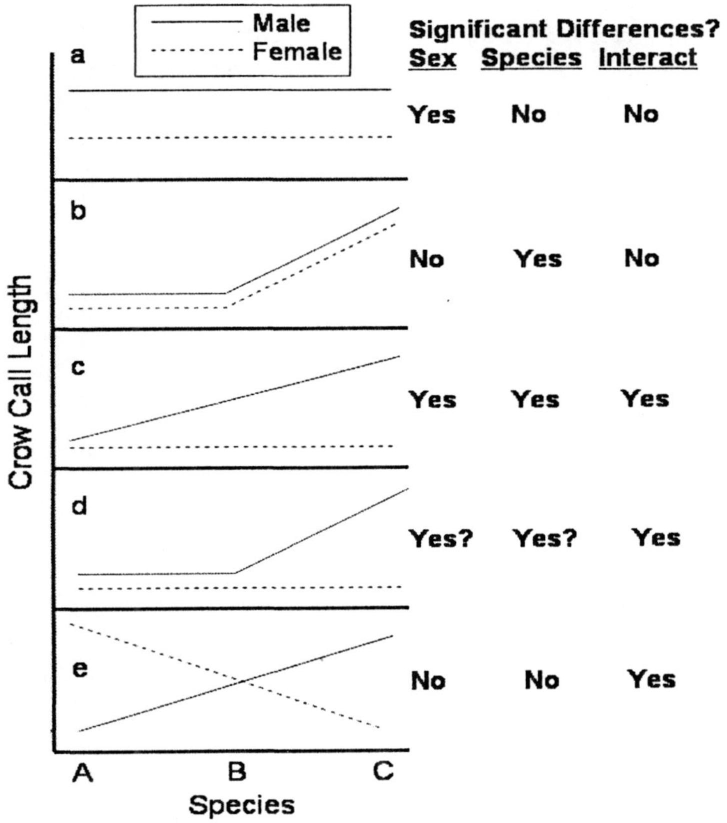

Main Effect: Influence of your independent variable on your dependent variable.

Interaction: The (possibly different) effect of one of your independent variables (or factors) at different levels of your other independent variable (or factor).

Linear or additive equation

There are multiple potential influences (e.g., species or sex) on the outcome for each score (e.g., an individual crow's call length) and those influences can be defined with a linear equation. While we will not directly use this formula in this chapter, we will discuss this a bit more in Chapter 11. For now it is important to focus on whether our treatment effects are enough to demonstrate a statistically significant difference between our groups. Conceptually, an individual's score (in this case, one crow's call length) could be represented in the following way:

$$\text{Score} = \mu + IV_1 + IV_2 + (IV_1)(IV_2) + E$$

Where:
μ = Combined influence of the grand mean
IV_1 = Effect of the first independent variable (sex)
IV_2 = Effect of the second independent variable (species)
$(IV_1)(IV_2)$ = Interaction effect of the independent variables (sex * species)
E = Unexplained error

What would the Data and the Analysis Look Like?

Notice in Table 9.1 that using two factors rather than one typically results in more groups, or more categories, into which subjects can be divided. If we were to break up our 15 subjects by sex as well species, it might look something like this. Obviously, this sample size would be considered extremely small for this analysis.

Table 9.1
Two-way ANOVA Sample Sizes (Crow Example)

		SPECIES		
		A	B	C
	Male	n = 2	n = 3	n = 3
SEX	Female	n = 3	n = 2	n = 2

In Table 9.2, we have provided a sample two-way ANOVA summary table. Both main effects (the row variable and the column variable) as well as the interaction are represented in the table. In this example, the row variable is sex and the column variable is species, but this would vary depending on the two independent variables you used in your study. The interaction is the combination of sex and species over and above any independent effect they might have on call length.

Table 9.2
Two-way ANOVA Summary Table (Crow Example)

Source	SS	df	S^2 (MS)	F obt.
Row (sex)				
Column (species)				
Row * Column (Sex * Species)				
Within (within a cell)				
Total				

s_w^2 = within-cell variance (reflects natural variation in the population, thus it is an estimate of σ^2). Within-cell variance could also be called "error" (E) because it is variability that we cannot control, but occurs naturally.
s_R^2 = between-row variance (reflects natural variation in the population AND any variability due to your row variable).
s_C^2 = between-column variance (reflects natural variation in the population AND any variability due to your column variable).
s_{RC}^2 = row * column variance (reflects natural variation in the population AND any variability due to the interaction between your row and column variables).

This two-factor ANOVA would produce three F-obtained values. One of them would be for a main effect of sex, one would be the main effect of species, and the last value would indicate whether or not there was a significant interaction of sex and species over and above their main effects (i.e., after accounting for the variance attributed to the main effects). These F values would be calculated by dividing the variance attributed to a main effect (i.e. row variance or s_R^2) by the within-cell variance (s_w^2), and then doing that again for the other main effect (i.e., column variance or s_c^2) and for the interaction term (s_{RC}^2). Mathematically, that would look like this:

F-obt for Row Factor (sex) = $\frac{s_R^2}{s_w^2}$

F-obt for Column Factor (species) = $\frac{s_C^2}{s_w^2}$

F-obt for Row*Column (interaction) = $\frac{s_{RC}^2}{s_w^2}$

Complete Example

Craik and Lockhart (1972) proposed a model of memory that suggested that the degree to which verbal material is remembered by the subject is a function of the degree to which it was processed when it was initially presented. To test this model, a researcher randomly assigned 50 subjects between the ages of 55 and 65 and a second group of 50 subjects between the ages of 21 and 25 to one of five groups: four incidental learning groups (without the expectation that the material will need to be recalled later) and one intentional-learning group. The incidental groups were asked to process the words in various ways, but were not told that they would later have to recreate the list of words. The intentional group was told to read through the list and to memorize the words for later recall. After subjects had gone through the list of 27 items three times, they were given a sheet of paper and asked to write down all of the words they could remember. The success of each group was compared using a two-way ANOVA. Age and memorization technique were the two factors (or independent variables), and the number of words from a list that were later recalled was the dependent variable. Figure 9.2 shows these data graphically. Complete the two-way ANOVA summary table based on the formulas for the three F-obtained values you were given above, and evaluate whether or not age and memorization techniques as well the interaction between them has a significant effect on the words correctly recalled. Assume that the data are normally distributed and the variances are homogeneous. Use $\alpha = 0.05$.

Null Hypothesis (Main Effect for Age): There is no difference in the number of words recalled based on age effects or the differences are due to chance.

> **Alternative Hypothesis (Main Effect for Age)**: There is a difference in the number of words recalled based on age effects.

Null Hypothesis (Main Effect for Technique): There is no difference in the number of words recalled based on technique or the differences are due to chance.

> **Alternative Hypothesis (Main Effect for Technique)**: There is a difference in the number of words recalled based on the technique employed.

Null Hypothesis (Interaction Age*Technique): There is no difference in the number of words recalled based on the interaction between age and technique or the differences are due to chance.

> **Alternative Hypothesis (Interaction Age*Technique)**: There is a difference in the number of words recalled based on the interaction between age and technique.

Figure 9.2

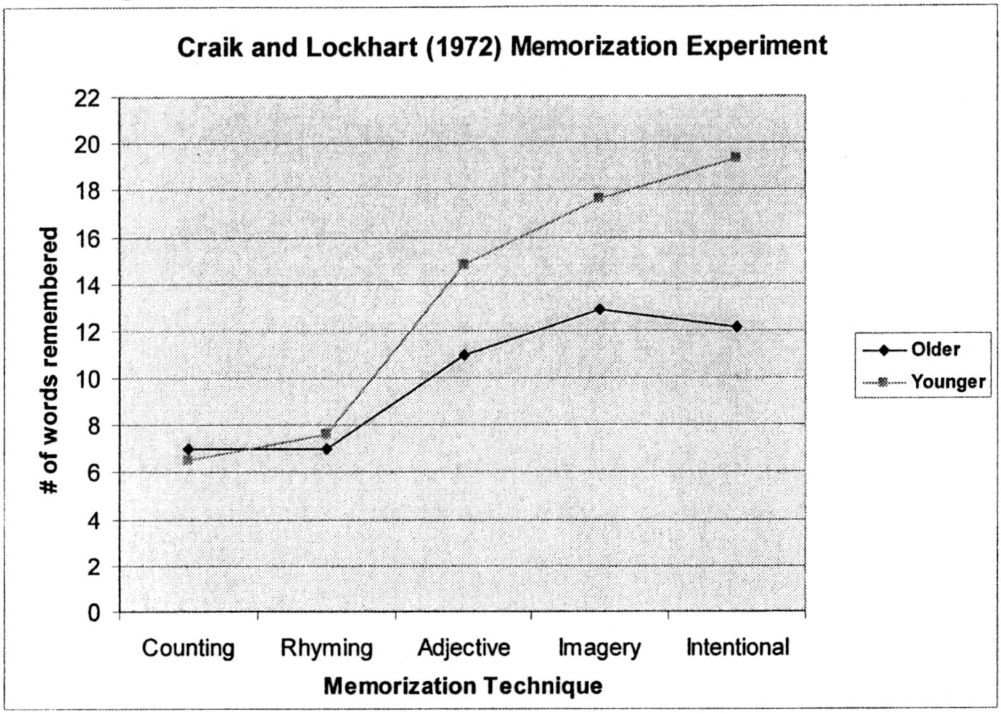

Table 9.3
Two-way ANOVA Summary Table (Memory Example)

Source	SS	df	S^2 (MS)	F obt.
Row (Age)	249.64	1		
Column (Technique)	1476.66	4		
Row * Column (Age * Technique)	195.26	4		
Within (within a cell)	628.80	90		
Total				

Step One: Apply the s^2 & F obtained formulas to complete the ANOVA summary table.

Calculate SS_T

$SS_T = SS_R + SS_C + SS_{RC} + SS_W$

$SS_T = 249.64 + 1476.66 + 195.26 + 628.80 = 2550.36$

Calculate df_T

$df_T = df_R + df_C + df_{RC} + df_W$

$df_T = 1 + 4 + 4 + 90 = 99$

Calculate Variances

Calculate variance the same way that you did in the one-way ANOVA, by dividing the SS by their appropriate df. If you have the table laid out properly, then you can divide the first column of numbers by their respective df (second column of numbers).

$$s_R^2 = \frac{SS_R}{df_R} = \frac{249.64}{1} = 249.64$$

$$s_C^2 = \frac{SS_C}{df_C} = \frac{1476.66}{4} = 369.165$$

$$s_{RC}^2 = \frac{SS_{RC}}{df_{RC}} = \frac{195.26}{4} = 48.815$$

$$s_w^2 = \frac{SS_W}{df_W} = \frac{628.80}{90} = 6.9833337$$

Calculate F-obtained values

$$\text{F-obt for Row Factor (Age)} = \frac{s_R^2}{s_w^2} = \frac{249.64}{6.9866667} = 35.73$$

$$\text{F-obt for Column Factor (Technique)} = \frac{s_C^2}{s_w^2} = \frac{369.165}{6.9866667} = 52.84$$

$$\text{F-obt for Row*Column (interaction)} = \frac{s_{RC}^2}{s_w^2} = \frac{48.815}{6.9866667} = 6.99$$

Step Two: Evaluating the probability of obtaining this score due to chance

Evaluate the F-obtained values based on alpha (α) = 0.05, and a two-tailed hypothesis. To evaluate your F-obtained value, you must use the F distribution (Table C). To determine your F-critical value, you need to know your alpha level (0.05) and your degrees of freedom (df_B and df_W). The degrees of freedom from the numerator of the F-obtained equation are the df_B while the degrees of freedom from the denominator are df_W. Thus, the df for this problem are 1 & 90, 4 & 90 and 4 & 90 respectively for the three F values you have calculated. Compare the appropriate F-critical value with the appropriate F-obtained value, and reject the null hypothesis when your F-obtained value is equal to or greater than your F-critical value. You should use 3.95, 2.47 and 2.47 as your F-critical values.

F obtained ≥ F critical in all three cases, so we *reject* the null hypotheses for both main effects (of age and technique) and we reject the null hypothesis for the interaction term.

How should we interpret these data in light of the effect of age and memorization technique on the ability to recall words from a list correctly?

These results suggest that age and technique have significant effects on the ability to recall a list of words, and that there is some interaction between age and technique that have effects above and beyond the influences of each separately. Based on the figure alone, it appears that younger adults recall more words than older adults, and that "adjective", "imagery" and "intentional techniques" are most successful. The "intentional" strategy is particularly successful in younger adults (interaction term). However, these conclusions are based purely on the appearance of the figure and knowledge that the overall F tests are significant. Proper multiple comparisons would need to be done to determine which cells are significantly different from which other cells. As we discussed in chapter 8, determining the statistical significance of these pairwise comparisons can only be done using "multiple comparisons" once you know that your overall test (your ANOVA) has reached statistical significance. Take a look at the Excel and SPSS output for these data that we have manually analyzed.

Results if you use Microsoft Excel to calculate the Two-way ANOVA

Anova: Two-Factor
ANOVA

Source of Variation	*SS*	*Df*	*MS*	*F*	*P-value*	*F crit*
Sample	249.64	1	**249.64**	**35.73091603**	4.4838E-08	**3.946865945**
Columns	1476.66	4	**369.165**	**52.83850191**	7.85493E-23	**2.472930305**
Interaction	195.26	4	**48.815**	**6.986879771**	6.04944E-05	**2.472930305**
Within	628.8	90	**6.9866667**			
Total	**2550.36**	**99**				

Note that Excel calls this test an "ANOVA: Two Factor". This wording is slightly different than we have been using, but it is describing exactly the same analysis. We have bolded the numbers that are comparable to the numbers we just manually calculated or looked up in the table. Recall that another name for variance is mean square (MS) and Excel uses a MS label in the table. Note that the F-obtained values (Excel calls it "F") are identical to ours, except they have carried more digits. The "F Crit" values are also the same, but again they have carried more digits. The degrees of freedom are the same. In addition, they have provided you with the exact probability ("P-value") of obtaining each F-obtained value ("F") of which is shown in scientific notation. The notation "4.4838E-08" tells us that we should move the decimal place left eight places, while "7.85493E-23" means that the decimal will have to be moved to the left 23 places. Thus, in all three cases, all of the probabilities are very, very small and definitely < 0.05, so we clearly *reject all* of our null hypotheses.

Results if you use SPSS to calculate the ANOVA (or F test)

Tests of Between-Subjects Effects

Dependent Variable: SCORE

Source	Type III Sum of Squares	df	Mean Square	F	Sig.
Model	15331.200	10	1533.120	219.435	.000
AGE	249.640	1	**249.640**	**35.731**	.000
TECH	1476.660	4	**369.165**	**52.839**	.000
AGE * TECH	195.260	4	**48.815**	**6.987**	.000
Error	628.800	90	**6.987**		
Total	15960.000	100			

The first F-value on the table is an overall F test that combines main effects and the interaction in one analysis. This is like the one-way ANOVA you did in chapter 8. The remaining F-values are just like the Excel printout except that there are no F-critical values provided and the p-value is now indicated as "Sig." If the "significance" value is less than your alpha level (which is usually 0.05), then you can reject the null hypothesis.

How would these results be reported in a scientific journal article?

If the journal required American Psychological Association (APA) format, the results would be reported in a format something like this:

> There was a significant difference in the number of words recalled based on age effects, $F(1,90) = 35.73, p < .01$.
>
> There was a significant difference in the number of words recalled based on the technique employed, $F(4,90) = 52.84, p < .01$.
>
> There was a significant interaction between age and technique that affected the number of words recalled, $F(4,90) = 6.99, p < .01$.

These formal sentences include the dependent variable (words recalled), the independent variables (age, technique and the interaction between age and technique), as well as a statement about statistical significance, the symbol of the test (F), the degrees of freedom between groups (1 or 4), the degrees of freedom within groups (90), the statistical values (35.73, 52.84 and 6.99), and the probability of obtaining these results simply due to chance (all <.01).

Another Complete Example

A researcher examined the outcomes of social interactions between two individuals, and hypothesized that the rate of very subtle eye glances towards the other interactor would predict the "winner" of the interaction. But first she wanted to answer a basic question: did the rates of eye glances differ based on the gender of the interactors and the type of discussion involved ("happy", "controversial" or "neutral" topics). The researcher placed independent male-male, male-female, and female-female pairs of individuals into one of the three discussion types, and coded the number of eye glances recorded in the session as a whole. The differences among the groups could be compared using a two-way ANOVA. Pair gender and discussion type were the two factors (or independent variables), and the number of eye glances was the dependent variable. Figure 9.3 shows these data graphically. Complete the two-way ANOVA summary table based on the formulas for the three F-obtained values you were given above, and evaluate whether or not pair gender and discussion type, as well the interaction between them, has a significant effect on the number of eye glances. Assume that the data are normally distributed and the variances are homogeneous. Use $\alpha = 0.05$.

Null Hypothesis (Main Effect for Pair Gender): There is no difference in the number of eye glances based on pair gender (male-male vs. male-female vs. female-female), or the differences are due to chance.

> **Alternative Hypothesis (Main Effect for Pair Gender)**: There is a difference in the number of eye glances due to the effects of pair gender.

Null Hypothesis (Main Effect for Discussion Type): There is no difference in the number of eye glances based on discussion type, or the differences are due to chance.

> **Alternative Hypothesis (Main Effect for Discussion Type)**: There is a difference in the number of eye glances due to the effects of discussion type.

Null Hypothesis (Interaction: Pair Gender by Discussion Type): There is no difference in the number of eye glances based on the interaction between pair gender and discussion type over and above the differences that are due to the main effects of pair gender and discussion type, or the differences are due to chance. There is an alternate form of this hypothesis: there is no difference in number of eye glances among the individual cell means, or the differences are due to chance.

Alternative Hypothesis (Interaction: Pair gender by Discussion Type): There is a difference in the number of eye glances due to the interaction between pair gender and discussion type. Alternatively, there is a difference among one or more of the cell means.

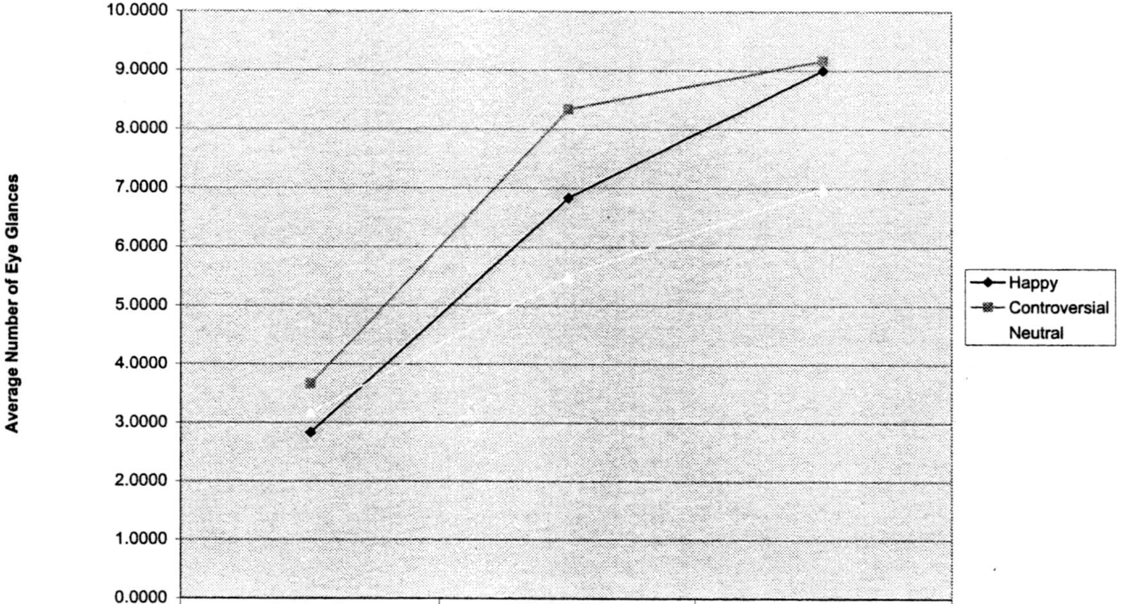

	Male-male	Male-female	Female-female
Happy	2.8333	6.8333	9.0000
Controversial	3.6667	8.3333	9.1667
Neutral	3.1667	5.5000	7.0000

Pair Gender

Table 9.3

Two-way ANOVA Summary Table (Eye glance example)

Source	SS	df	S^2 (MS)	F obt.
Row (Discussion Type)	30.33333	2		
Column (Pair Gender)	254.3333	2		
Row * Column (Discussion Type * Pair Gender)	13.33333	4		
Within (within a cell)	349.5	45		
Total				

Step One: Apply the s^2 & F obtained formulas to complete the ANOVA summary table.

Calculate SS_T

$SS_T = SS_R + SS_C + SS_{RC} + SS_W$

$SS_T = 249.64 + 1476.66 + 195.26 + 628.80 = 2550.36$

Calculate df_T

$df_T = df_R + df_C + df_{RC} + df_W$

$df_T = 1 + 4 + 4 + 90 = 99$

Calculate Variances

Calculate variance the same way that you did in the one-way ANOVA, by dividing the SS by their appropriate df. If you have the table laid out properly, then you can divide the first column of numbers by their respective df (second column of numbers).

$$s_R^2 = \frac{SS_R}{df_R} = \frac{30.3333}{2} = 15.1667$$

$$s_C^2 = \frac{SS_C}{df_C} = \frac{254.3333}{2} = 127.1667$$

$$s_{RC}^2 = \frac{SS_{RC}}{df_{RC}} = \frac{13.33333}{4} = 3.3333$$

$$s_w^2 = \frac{SS_W}{df_W} = \frac{349.5000}{45} = 7.7667$$

Calculate F-obtained values

F-obt for Row Factor (Discussion Type) $= \frac{s_R^2}{s_w^2} = \frac{15.1667}{7.7667} = 1.9528$

F-obt for Column Factor (Pair Gender) $= \frac{s_C^2}{s_w^2} = \frac{127.1667}{7.667} = 16.3734$

F-obt for Row*Column (interaction) $= \frac{s_{RC}^2}{s_w^2} = \frac{3.3333}{7.7667} = 0.4292$

Step Two: Evaluating the probability of obtaining this score due to chance

Evaluate the F-obtained values based on alpha (α) = 0.05, and a two-tailed hypothesis. To evaluate your F-obtained value, you must use the F distribution (Table C). To determine your F-critical value, you need to know your alpha level (0.05) and your degrees of freedom (df_B and df_W). The degrees of freedom from the numerator of the F-obtained equation are the df_B while the degrees of freedom from the denominator are df_W. Thus, the df for this problem are 2 & 45, 2 & 45 and 4 & 45, respectively for the three F values you have calculated. Compare the appropriate F-critical value with the appropriate F-obtained value, and reject the null hypothesis when your F-obtained value is equal to or greater than your F-critical value. You should use 3.21, 3.21 and 2.58 as your F-critical values.

F obtained ≥ F critical only in the case of Pair Gender, so we *reject* the null hypothesis for the main effect of Pair Gender and *fail-to-reject* the null hypotheses for the main effect of Discussion Type and interaction.

How should we interpret these data in light of the effect of pair gender and discussion type on the number of eye glances?

These results suggest that pair gender has a significant effect on the number of eye glances, as can be seen in the figure: male-male pairs have relatively fewer eye glances, regardless of the discussion type, while male-female and female-female pairs exhibit increasing numbers of eye glances, again regardless of the discussion type. When discussion type alone is considered, there is no significant difference in number of eye glances, and there is no interactive effect of pair gender and discussion type, over and above the main effects already described. However, the direction of the pair gender effect is based entirely on the appearance of the figure and the knowledge that the overall F test for pair gender is significant. Proper multiple comparisons would need to be done to determine which cells are significantly different from which other cells. As we discussed in chapter 8, determining the statistical significance of these pairwise comparisons can only be done using "multiple comparisons" once you know that your overall test (your ANOVA) has reached statistical significance. Take a look at the Excel and SPSS output for these data that we have manually analyzed.

Results if you use Microsoft Excel to calculate the Two-way ANOVA

ANOVA

Source of Variation	*SS*	*df*	*MS*	*F*	*P-value*	*F crit*
Sample	30.33333	2	**15.16667**	**1.95279**	0.153715	**3.20432**
Columns	254.3333	2	**127.1667**	**16.37339**	4.54E-06	**3.20432**
Interaction	13.33333	4	**3.333333**	**0.429185**	0.786769	**2.578737**
Within	349.5	45	**7.766667**			
Total	**647.5**	**53**				

We have bolded the numbers that are comparable to the numbers we just manually calculated or looked up in the table. Recall that another name for variance is mean square (MS) and Excel uses a MS label in the table. Note that the F-obtained values (Excel calls it "F") are identical to ours, except they have carried more digits. The "F Crit" values are also the same, but again they have carried more digits. The degrees of freedom are the same. In addition, they have provided you with the exact probability ("P-value") of obtaining each F-obtained value ("F") of which is shown in scientific notation. The notation "4.54E-06" tells us that we should move the decimal place left six places (0.00000454). Thus, only in the case of the pair gender effect is the probability small,i.e., < 0.05, and so we only *reject* that null hypothesis.

Results if you use SPSS to calculate the ANOVA (or F test)

Tests of Between-Subjects Effects

Dependent Variable: EYEGLAN

Source	Type III Sum of Squares	df	Mean Square	F	Sig.
Model	298.000	8	37.250	4.796	.000
DISCTYP	30.333	2	**15.167**	**1.953**	.154
PAIRGEN	254.333	2	**127.167**	**16.373**	.000
DISCTYP * PAIRGEN	13.333	4	**3.333**	**.429**	.787
Error	349.500	45	**7.767**		
Total	2701.000	54			

The first F-value on the table is an overall F test that combines main effects and the interaction in one analysis. This is like the one-way ANOVA you did in chapter 8. The remaining F-values are just like the Excel printout except that there are no F-critical values provided and the p-value is now indicated as "Sig." If the "significance" value is less than your alpha level (which is usually 0.05), then you can reject the null hypothesis.

How would these results be reported in a scientific journal article?

If the journal required American Psychological Association (APA) format, the results would be reported in a format something like this:

> There was a significant difference in the number of eye glances based on pair gender (male-male vs. male-female vs. female-female), $F(2,45) = 16.37$, $p < .01$.
>
> There was no significant difference in the number of eye glances due to the effects of discussion type, $F(2,45) = 1.95$, $p = .15$.
>
> There was no significant difference in the number of eye glances based on the interaction between pair gender and discussion type over and above the differences that are due to the main effects of pair gender and discussion type, $F(4,45) = .43$, $p = .79$.

These formal sentences include the dependent variable (eye glances), the independent variables (gender pairing, discussion type and the interaction between gender pairing and discussion type), as well as a statement about statistical significance, the symbol of the test (F), the degrees of freedom between groups (2 or 4), the degrees of freedom within groups (45), the statistical values (16.37, 1.95 and .43), and the probability of obtaining these results simply due to chance (<.01, .15 and .79 respectively).

When is it appropriate to use a Two-way ANOVA test?

1) Use a Two-way ANOVA when you have two independent variables and a between-groups (independent groups) design, and
2) When the sampling distribution is normally distributed. Again, this is satisfied when:
   a. The sample size is greater or equal to 30 in each group ($n \geq 30$) or,
   b. The null hypothesis population is known to be normally distributed.

   and
5) When the dependent variable is on an interval or ratio scale (although some statisticians believe that this is too strict), and
6) When the variances of the groups are the same, or are **homogeneous**. The homogeneity of variance assumption (HOV) requires that the variances of the samples are equal, or in practical terms, not significantly different from one another. As before, we will use a simple rule of thumb to decide whether the variances are similar enough to be considered homogeneous:

   Rule of Thumb for HOV check: if the $\frac{\text{LargestVariance}}{\text{SmallestVariance}} \leq 4$

   then HOV assumption is met (okay) and you can proceed with ANOVA. If not, you cannot proceed with the ANOVA (we'll explain what to do instead later).

Multifactorial ANOVA

Multifactorial ANOVA is simply the situation where you have even more than two independent variables. Three-way, Four-way and more-way ANOVA's are conceptually possible, although the interaction terms become more and more difficult to interpret. This is because you can have two-way and three-way interactions in the same three-way ANOVA analysis. You will also hear the term "MANOVA" which describes Multiple ANOVA, or the situation where you have multiple dependent variables, not independent variables. This topic is beyond the scope of this textbook, but it is important to keep analyses with multiple independent variables conceptually separate from analyses with multiple dependent variables.

Linear or additive equation applied to the Multifactorial ANOVA

There are multiple potential influences (represented by each of your independent variables) on the outcome for each score (or value of your dependent variable) and those influences can be defined with a linear equation. If you are doing a three-way ANOVA, then an individual's score could be represented in the following way:

$$Score = \mu + IV_1 + IV_2 + IV_3 + (IV_1)(IV_2) + (IV_1)(IV_3) + (IV_2)(IV_3) + (IV_1)(IV_2)(IV_3) + E$$

Where
μ = Combined influence of the grand mean
IV_1 = Effect of the first independent variable
IV_2 = Effect of the second independent variable
IV_3 = Effect of the third independent variable
$(IV_1)(IV_2)$ = Interaction effect of the first two independent variables
$(IV_1)(IV_3)$ = Interaction effect of the first and third independent variables
$(IV_2)(IV_3)$ = Interaction effect of the second and third independent variables
$(IV_1)(IV_2)(IV_3)$ = Interaction effect of all three independent variables
E = Unexplained error

Repeated Measures ANOVA

A with-in groups or repeated-measures ANOVA is similar to a paired t-test except that you have more than two levels of your independent variable. In other words, you have a within-groups design (subjects are repeated or paired/matched), but your independent variable might include three or more levels. For example, let's say that you want to examine the effect of age-related changes on attention span and you measure the same 50 subjects three times. Twenty-five of the subjects are male and twenty-five are female. On the first occasion that you bring them into the laboratory, they are 3 years old, the second time they are 5 years old, and the last time they are 7 years old. Since the same subjects were used in all three "treatments", we can estimate error or variance due to differences among subjects (Subj) alone. Thus, each score is the result of a combined influence of the grand mean (μ), any treatment effect attributable to our independent variable (IV = Gender), any effect attributable to individual differences between subjects (Subj) and unexplained error (E).

Linear or additive equation

Score = $\mu + IV_1 + Subj + E$

In the same way that the paired t-test was inherently more powerful than the independent t-test, the repeated-measures one-way ANOVA is typically more powerful than the independent one-way ANOVA. This is because we have "removed" or "explained" more of the "unexplained" variance. Specifically, we have removed the portion of the variance that was due to individual or subject differences. We can calculate those differences now because we have multiple measures on the same subjects. It is quite apparent when a subject is consistently above or below the mean, and this variability can be quantified. The remaining unexplained variance (or error) continues to be the number used in the denominator to calculate the F-obtained value. Having a smaller denominator (because we have reduced the unexplained error) results in a larger F-obtained value, making it more likely that we can reject the null hypothesis when it is actually false. We have increased power! Repeated-measures ANOVA is similar to the one-way independent ANOVA from chapter 8 because you do not calculate an interaction between your independent variable and effect of subject (Subj).

When is it appropriate to use a Repeated Measures ANOVA?

1) Use a repeated-measures ANOVA when you have one independent variable with more than two levels and a within-groups design, and
2) When the sampling distribution is normally distributed. Again, this is satisfied when:
   a. The sample size is greater or equal to 30 in each group ($n \geq 30$) or,
   b. The null hypothesis population is known to be normally distributed.

and

3) When the dependent variable is on an interval or ratio scale (although some statisticians believe that this is too strict).

Overview of Single Sample, Two Sample & 3+ Sample Tests: Expanded Flowchart

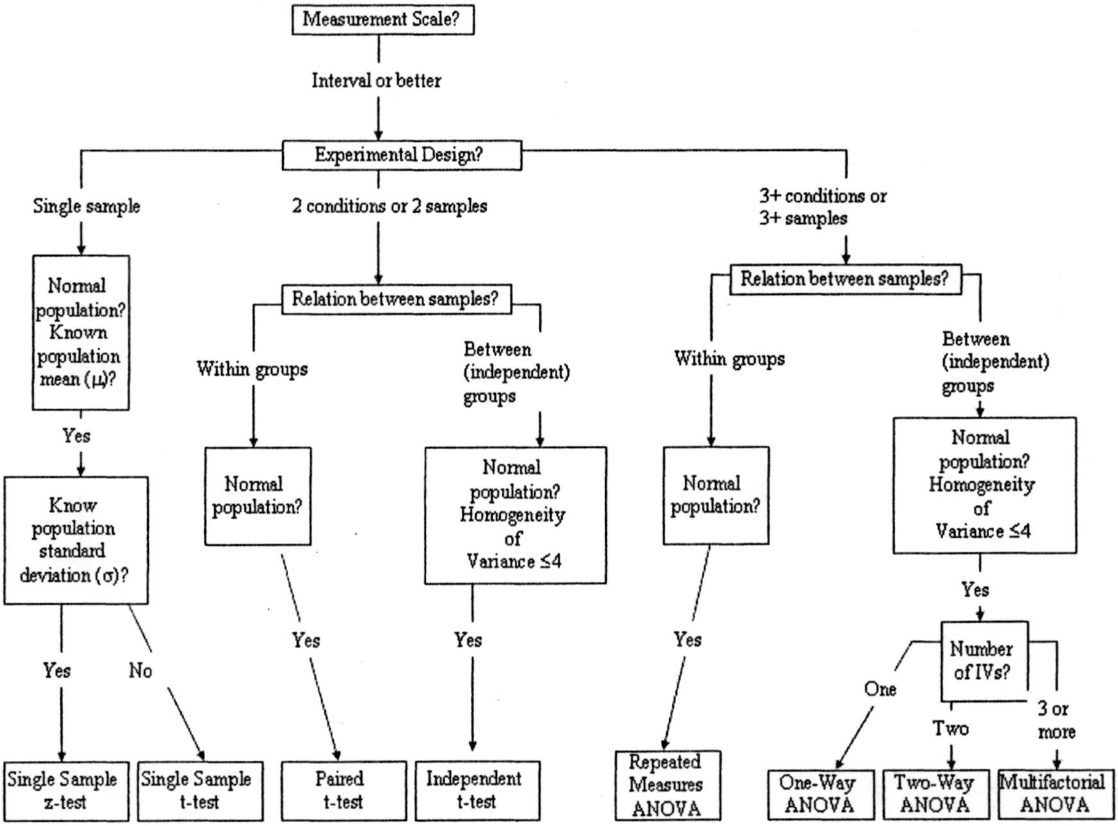

Notice that the flowchart does not yet tell you what to do if you do not meet the assumptions of your statistical tests. In fact, there are some options when you have three samples, but you do not meet the assumptions for parametric tests. One option is to perform a log transformation on your data to determine if the mathematical transformation will result in a normal distribution.

Summary

In this chapter we have focused on particular types of ANOVA where you have a within-groups design or a between-groups (independent) design with multiple independent variables. Despite our coverage here, there are many more possibilities including within-groups designs with more than one independent variable and situations where some of the data that are collected are interval or ratio, but other variables you want to include are not interval or ratio, etc. We'll introduce some other powerful approaches using ANOVA in chapter 12.

Chapter Nine Homework

Provide a short answer for the following questions.

1. How does a repeated measures ANOVA differ from a one-way ANOVA?
2. What are the assumptions for a two-way ANOVA?
3. What are the assumptions for a repeated measures ANOVA?
4. How does a two-way ANOVA differ from a one-way ANOVA?
5. What does the linear equation (presented twice in this chapter) help us to understand?
6. Why is the repeated measures ANOVA more powerful than the one-way (independent) ANOVA?
7. What is a main effect?
8. What is an interaction effect?
9. Does the repeated-measures ANOVA have an interaction term? Why or why not?

Complete the ANOVA summary tables for problems 10 – 13.

10. You have just completed a study with three treatment groups (low, medium and high). There were ten males and ten females (or 20 participants) in each of the three treatment groups. Complete the ANOVA source table provided below. Using an $\alpha = 0.05$, what do you conclude? Reject or Fail-to-Reject the H_0 for each F obtained value.

ANOVA

Source of Variation	*SS*	*df*	*MS*	*F*	*F crit*
Row (sex)	828.8167	1			
Columns (treatment)	1265.633	2			
Interaction (sex * treatment)	847.2333	2			
Within	14768.9	54			
Total					

11. You have just been contacted by a colleague that needs help with analyzing a data set. Their study had four treatment groups (Married, Divorced, Single, and Separated). There were eight males and eight females (or 16 participants) in each of the four treatment groups. Complete the ANOVA source table provided below. Using an $\alpha = 0.05$, what do you conclude? Reject or Fail-to-Reject the H_0 for each F-obtained value.

ANOVA

Source of Variation	*SS*	*df*	*MS*	*F*	*F crit*
Row (sex)	495.0625	1			
Columns (status)	1791.625	3			
Interaction (sex * status)	1855.063	3			
Within	15426.25	56			
Total					

12. An organizational psychologist evaluated the effects of job category and blood pressure on the amount of fines those individuals had paid on parking and traffic tickets in the last ten years. The two job categories were lawyer and doctor. There were four categories of blood pressure (optimal/below normal, normal, high normal and very high). They had six lawyers and six doctors in each of the four categories of blood pressure that they evaluated for parking/traffic fines. They hypothesized that individuals with higher blood pressure were more likely to be stressed and thus more likely to be in a hurry and be fined for illegal parking or moving violations. Both job categories were high on income, so the penalty for parking/moving violations would not be financially difficult in and of itself. Complete the ANOVA source table below and interpret the outcome of the study based on the hypothesis and whether or not you can reject the null hypothesis for each F value.

ANOVA

Source of Variation	*SS*	*df*	*MS*	*F*	*F crit*
Row (job)	560952.5	1			
Columns (blood pressure)	10028590	3			
Interaction (job * blood pressure)	199493.2	3			
Within	10005058	40			
Total					

CHOOSE THE MOST APPROPRIATE AND POWERFUL TEST:

13. Five different elementary schools were selected to participate in a study on reading comprehension. The 5th grade classes at each school were assessed for reading comprehension using a standardized exam at the beginning of the study, and there was no significant difference found between the groups. Four different reading programs were used for the next six months and one school continued to teach their students using the techniques that had used previously (control group). After the six month period, the reading comprehension scores of 5th graders at the five schools were again compared to one another with the following results:

Treatment A	Treatment B	Treatment C	Treatment D	Control
Mean = 45 N = 60	Mean = 55 N = 50	Mean = 40 N = 77	Mean = 60 N = 68	Mean = 40 N = 54

Variances between the groups were homogeneous. What is the appropriate analysis for these results? Why is that the correct analysis?

14. Two groups of seniors with Alzheimer's disease were compared on their ability to remember the details of an event that occurred in the previous week. Memory was measured as the number of specific

details they were able to remember from the event. If no correct details were remembered they received a score of zero. The scores ranged from 0 – 19. The first group (n = 30) was receiving drug A and the second group (n = 44) was receiving drug B. The variances were homogeneous between the two groups. What is the proper analysis to compare the memory capabilities of the two groups based on their drug treatment and why?

15. A cognitive psychologist exploring normal memory decay brought forty individuals into the laboratory to "plant" a specific memory. This memory was the exact details of the laboratory experience. The forty individuals were brought back one month later, six months later, one year later, and three years later to be assessed on their ability to remember the details of their first experience. Five subjects failed to appear for the last assessment, and were dropped from the study and analysis. Memory was recorded as the number of specific (and correct) details recalled about the event. Each error reduced their score by one digit. Scores ranged from -3 to +12. The variances between groups were homogeneous. What is the proper analysis to use to compare their memories at each of the laboratory visits?

Chapter 10
Regression and Correlation

In this chapter, we discuss linear relationships between two variables. In this case, both the independent and the dependent variables are measured on an interval or ratio scale, and the variables are typically referred to as X and Y. In previous chapters, the independent variable has always been categorical (e.g., male vs. female or low dose, high dose, and control dose), but that is not always the case and we need to perform the appropriate analyses when our independent variable is continuous rather than categorical. An example will help illustrate the concept.

Linear Regression

General Overview

Suppose that we examine the relationship between the number of hours a week that ten students spend on studying and their cumulative grade point average (GPA; Table 10.1). Both variables are continuous. The independent variable is the number of hours of effort and the dependent variable is GPA. We can graph those results as a scatterplot in Microsoft Excel (Figure 10.1).

Student	Hours (X)	Grade (Y)
1	10	3.4
2	12	3.5
3	5	2.7
4	2	1.7
5	20	3.9
6	14	3.6
7	11	3.3
8	22	3.8
9	3	1.9
10	18	4

Figure 10.1

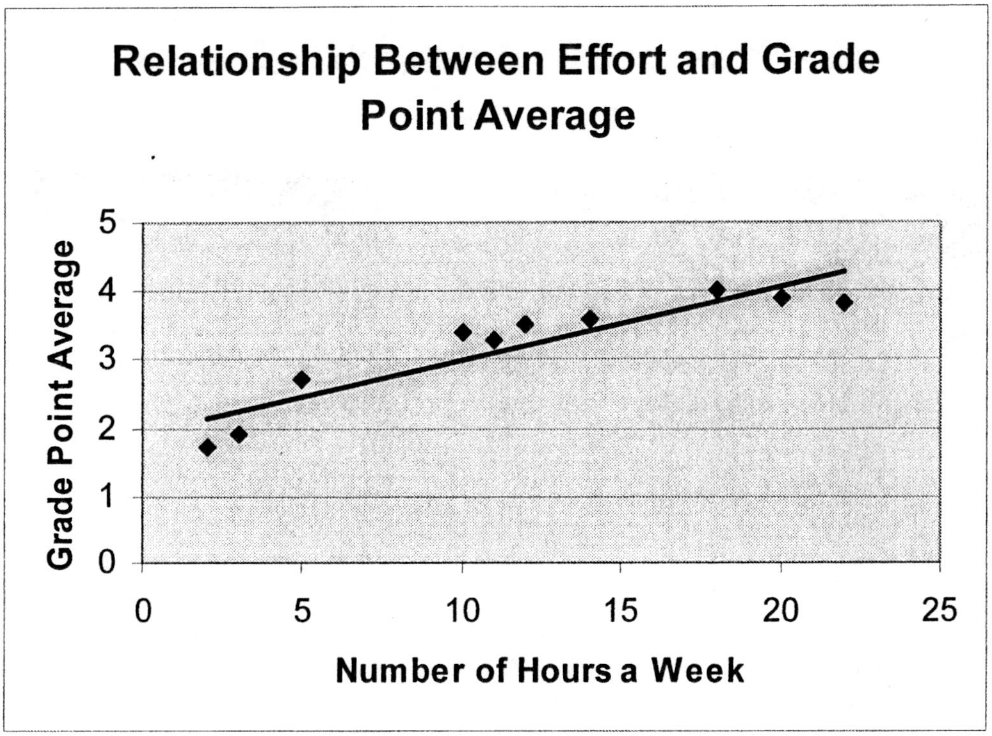

The paired data points from all of our ten students are shown clearly in Figure 10.1. Note that our independent variable (or X variable) is shown on the horizontal or X-axis, while our dependent variable (Y) is shown on the vertical or Y-axis. For each student, we went over X units and up by Y units to plot their data. If you do this for all of your data points, you can get a visual estimate of the overall relationship between X and Y. In this case, it appears that as X increases, Y increases. In fact, there appears to be a linear relationship between X and Y. We could draw a straight line that would approximate the relationship between our two variables, essentially by creating a line that is the "average" of all of the data (Figure 10.2). This is a lot like the concept of a grand mean that was introduced in the ANOVA chapter.

Figure 10.2

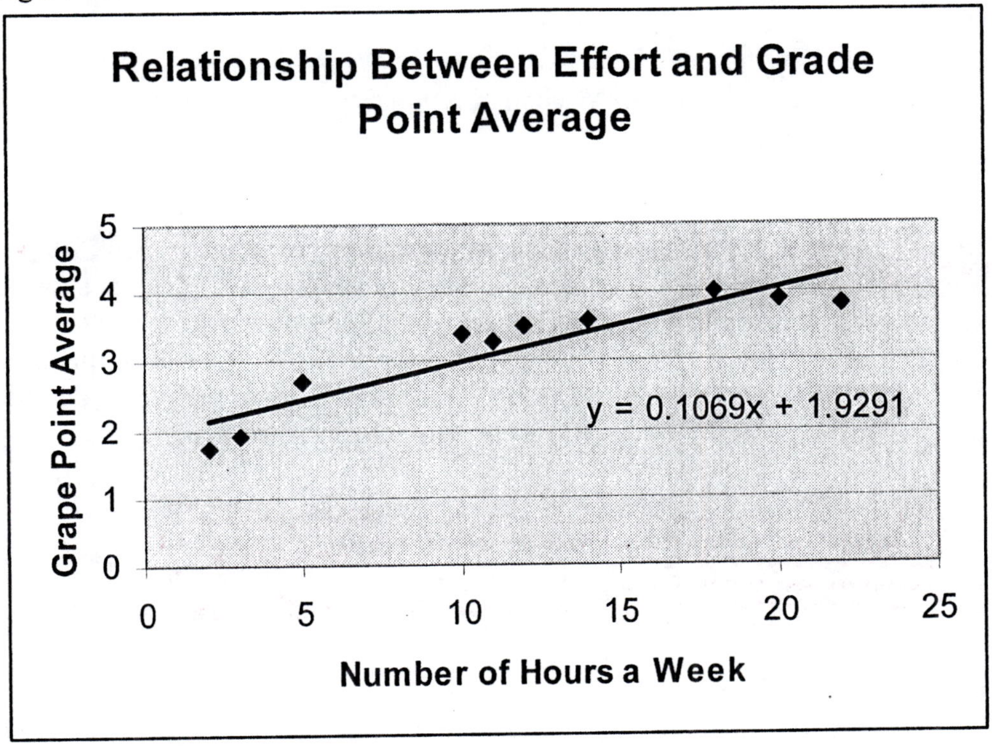

The line that we have drawn can be expressed mathematically. It is the "best fit" line and is expressed in the general form:

$Y = bX + a$ (this is the form that Excel has used and is shown in Figure 10.2)

or also commonly reversed:

$Y = a + bX$

where
X = X score,
Y = Y score,
a = Y intercept, or the value of Y when X is equal to zero, the point at which the line crosses the X-axis,

b = slope of the line = $\frac{\Delta Y}{\Delta X}$ or $\frac{rise}{run}$

Y intercept: The value of Y when X is equal to zero, which is where the line crosses the Y-axis.
Slope: The change in Y divided by the change in X.

In this example, the relationship between your X and Y variables is **positive (direct)**. The alternative would be a **negative** or **inverse** relationship, such as the one you would expect between the amount of partying and the GPA (higher level of partying is associated with a lower GPA). This example is NOT an example of a **perfect** relationship because many of the points do not fall exactly on the best-fit line, so it is an **imperfect** relationship.

Positive (Direct) Relationships: As X increases, Y increases.
Negative (Inverse) Relationships: As X increases, Y decreases.
Perfect Relationships: All data points fall on the best-fit line.
Imperfect Relationships: All data points do not fall on the best-fit line.

Applications of Linear Regression

Linear regression is a technique that uses the relationship between two variables for prediction. This implies a causal relationship between your two variables. If this is a safe assumption, you can then use the equation of the line to predict Y values for any given X (within the boundaries of your range of X scores). The best fit or "least squares" regression line is the line that minimizes the squared differences between your actual Y and your predicted Y values (Y'). The slope and the y-intercept are calculated using the least squares method that you learned when we talking about calculating a mean in chapter 3. In fact, you can think of the line as the "mean" for each X value. With minor modifications of the general form of the linear equation,

$Y = bX + a$

we can create the Least Squares Regression Line:

$Y' = b_yX + a_y$

where:
Y' = predicted value of Y
b_y = slope of the line that minimizes the errors in predicting Y from X
a_y = y-intercept of the line that minimizes the errors in predicting Y from X

Equation of the line for our Example Data:

You can produce the best-fit line equation for your data directly from Excel, which is shown in Figure 10.2, but, in a sidebar, we'll also show you how to calculate these numbers on your own.

$Y' = 0.1069(X) + 1.9291$

One you have this formula, you can use it to predict values of Y given an X of your choice. For example, you could predict the GPA of a student who works on homework/studying approximately 13 hours a week.

Equation of the line applied to a new value of X:

$Y' = 0.1069(X) + 1.9291$
$Y' = 0.1069(13) + 1.9291 = 3.3188$

Based on the information provided in this dataset, your best estimate of the GPA of a student who studied 13 hours a week would be 3.3.

Standard Error of the Estimate

While we have succeeded in estimating GPA from effort, we need to know the variability around our estimate. If the least squares regression line can be thought of as a "mean", then what is the "standard deviation"? In this case, we use the "standard error of the estimate". This is the amount of error around the estimate, just like the standard deviation measures error around the mean.

Because the formula for calculating SEE is cumbersome, yet the concept is similar to what you have already done, we will provide the SEE in the example and exercise problems for you and focus on interpretation instead of manual calculation.

For the example on studying and GPA, the SEE = 0.347858417. This means that the variability of the points around the regression line is approximately 0.35 grade point units, so the overall ability to predict GPA from effort spent in studying is good, but not great. Indeed, the larger your SEE, the less confident we can be in any of the predications of Y (Y').

When is it appropriate to use a Regression analysis?

1) Use a regression when you have an independent and a dependent variable on an interval or ratio scale, and
2) When the relationship between the two variables are linear (rather than curvilinear, etc.), and
3) When your data are **homoscedastic**. This means that the variability around the regression line is uniform for all of the values of X (see Figure 10.3).

Figure 10.3

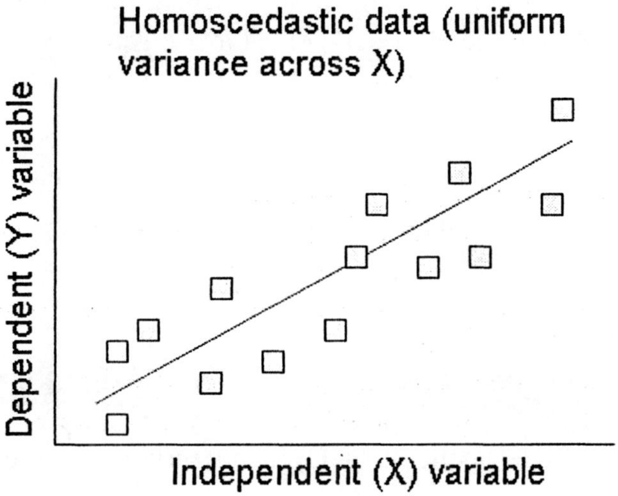

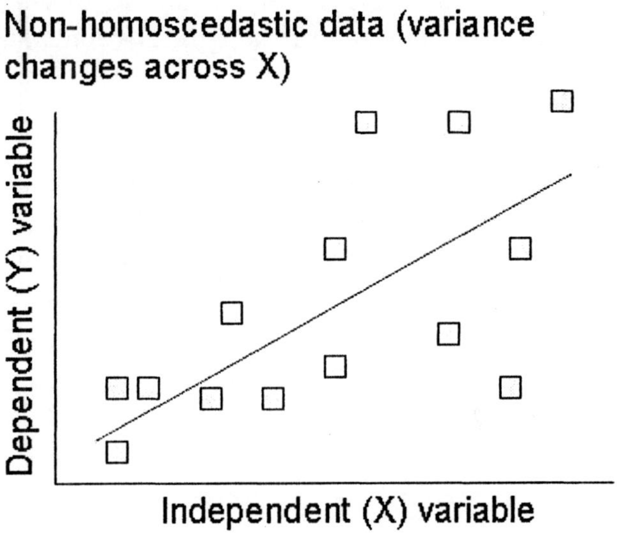

Homoscedastic: variance around line is uniform across all X's.

Linear or additive equation

Each score (Y) is the result of the combined influence of the X variable and error (E or SEE). We can add the error component to the equation for the regression line that we developed earlier in the chapter:

$$Y = a + b(X) + E$$

This looks similar to the linear equation that we developed for ANOVA, and indeed they are related (see chapter 11).

Multiple Regression

Multiple regression is like the two-way ANOVA because it has more than one independent variable effect that is assessed at the same time. If we were to add a measure of intelligence quotient (IQ) to our earlier example, we could look at the independent effects of both IQ and effort (studying) on GPA. We will show the SPSS out put for a multiple regression later in the chapter, but we need to discuss correlation analysis before you will fully understand that output.

Linear or additive equation for multiple regression

Conceptually, each score (Y) is the result of the combined influence of each X variable and error (E or SEE). We can add the second independent variable component to the equation for the simple regression line that we just discussed:

$$Y = a + b_1(X_1) + b_2(X_2) + E$$

Correlation

A correlation is a statistic that expresses the degree of the fit of the data to a line, and the type of relationship (direct or inverse). Unlike regression techniques, correlation coefficients make no assumptions about the cause-and-effect direction of an X-Y relationship, but rather simply measure the degree to which two sets of paired scores vary together in a consistent

(linear) manner. Thus, you can calculate a correlation coefficient for regression-type "cause-and-effect" data, but you cannot necessarily calculate a "cause-and-effect" regression line for correlational data.
Correlation coefficients have the following characteristics:

1. Values always range between -1 and +1
2. A positive coefficient indicates a direct relationship (positive slope), whereas a negative coefficient (negative slope) indicates an inverse relationship.
3. A coefficient that is equal to zero indicates that there is no relationship between the two variables (supports your null hypothesis).
4. A coefficient that is equal to -1.00 indicates that you have a perfect inverse relationship between your variables, while a coefficient that is equal to +1.00 indicates that there is a perfect direct (positive) relationship between your variables.

Figure 10.4

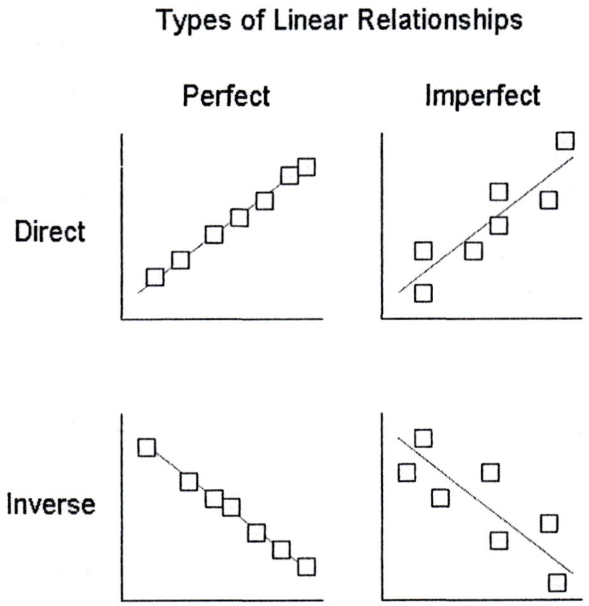

Pearson's r Correlation Coefficient

The most common correlation coefficient is the "Pearson's r" or "Pearson Product-Moment Correlation Coefficient", but there are many other correlation coefficients to choose from. We will discuss a few more later in the chapter. The Pearson's r requires that your data (independent and dependent variables) be interval or ratio scale. The Pearson r is a measure of the extent to which your paired scores occupy the same or opposite relative positions in their own distribution. Intuitively, that suggests that lower X values are associated with lower Y values (direct relationship), or that lower X values are associated with higher Y values (inverse relationship).

Interpreting the Correlation Coefficient

The Pearson's r coefficient for our study time-GPA example data is 0.91. A correlation coefficient this close to perfect (near one) clearly indicates a strong relationship between our two variables. But what we really need to do is to determine the probability that we could get a statistic of 0.91 or more extreme by chance, the same question that we have been asking throughout much of Unit 2 and 3. For the Pearson's r, we are determining the probability that, given a sample coefficient of 0.91, the correlation coefficient of the underlying population is really zero, reflecting no relationship between X and Y. Like our earlier statistical tests, there is a table for critical values of r based on the degrees of freedom (N-2) and the desired level of alpha (see Table E). Even though you have shown a high correlation between the time spent studying each week and grade point average, you cannot demonstrate a causal link between the two variables with correlational results. It is possible that some third factor that you didn't measure (such as IQ) is really driving both of your variables independently from one another.

Another way to express the relationship between the independent and the dependent variables is to square your r-obtained value to make it R^2.

r =

$$\sqrt{proportion - of - total - \text{var}iance - of - Y - that - is - \text{exp}lained - by - X}$$

R^2 = The proportion of the variability in Y that is explained by X.

The use of the capitalized R is intentional as R^2 is the formal symbol for this statistic. If we apply this concept to our example, then R^2 is the proportion of the variance in grade point average that is explained by how much time students spent studying. Obviously, there are other possible factors that influence grade point average, such as how hard you study, your natural aptitude, and the quality of the instruction at your college. If $r = 0.9154296$, then $R^2 = (0.9154296)^2 = 0.838011$. This means that approximately 84% of the variability in grade point average is explained by how much time students spend studying. What if your correlation coefficient is much smaller, so that $r = 0.30$?

$R^2 = (0.30)\ (0.30) = 0.09$

$0.09 \text{ X } 100 = 9\%$

If this were the case, then only 9% of the variability in your dependent variable would be explained by your independent variable, with 91% of the variability in your dependent variable (Y) left unexplained. This would suggest that the factor you are studying does not have a strong relationship with your dependent variable.

Results if you use Microsoft Excel to calculate a regression on study time-GPA data

SUMMARY OUTPUT

Regression Statistics	
Multiple R	**0.91542963**
R Square	**0.838011408**
Adjusted R Square	0.817762834
Standard Error	**0.347858417**
Observations	**10**

ANOVA

	df	*SS*	*MS*	*F*	*Significance F*
Regression	1	5.007956174	5.007956174	41.38619382	0.000201874
Residual	8	0.968043826	0.121005478		
Total	9	5.976			

	Coefficients	*Standard Error*	*t Stat*	*P-value*	*Lower 95%*	*Upper 95%*
Intercept	**1.929080119**	0.223406068	8.634859975	2.5101E-05	1.413904468	2.44425577
Hours(X)	**0.106916229**	0.016619421	6.433210227	0.000201874	0.068591751	0.145240708

This is the Excel output for the study time and GPA data. We have bolded the numbers that are comparable to the numbers that we just manually calculated or that you were given in the problem (SEE). Note that the F-obtained value (Excel calls it "F") is testing whether we have a significant overall relationship between X and Y. In addition, they have provided you with the exact probability ("P-value") of obtaining an F-obtained value ("F") of 41.38619382 which is 0.000201874. These results suggest that exactly 0.0201874% of the time, you would obtain this measure of association between X and Y simply due to chance. Thus, the likelihood that the association is due to chance is not high, but you still do not know whether X causes Y.

Results if you use SPSS to calculate the regression

Model Summary

Model	**R**	**R Square**	Adjusted R Square	**Std. Error of the Estimate**
1	**.915**	**.838**	.818	**.3479**

a Predictors: (Constant), HOURS

ANOVA

Model		Sum of Squares	df	Mean Square	F	Sig.
1	Regression	5.008	1	5.008	41.386	.000
	Residual	.968	8	.121		
	Total	5.976	9			

a Predictors: (Constant), HOURS
b Dependent Variable: GPA

Coefficients

Model		Unstandardized Coefficients B	Std. Error	Standardized Coefficients Beta	t	Sig.
1	(Constant)	**1.929**	.223		8.635	.000
	HOURS	**.107**	.017	.915	6.433	.000

a Dependent Variable: GPA

The output from SPSS is similar to the output from Excel, but the Y-intercept and the slope are less clearly labeled. They appear in the last table marked "Coefficients" and are highlighted. Note that both Excel and SPSS have given us our r-obtained value along with our regression results. Let's look next at what you get when you actually ask for a correlation analysis.

Results when you use Microsoft Excel to calculate a Correlation

	Hours(X)	*Grade(Y)*
Hours(X)	1	
Grade(Y)	**0.91543**	1

Excel gives you a very simple table that contains your X and Y variables and the correlation between Hours and Grade (0.91543). The ones in the table are the correlation between hours with hours and grade with grade, which is 100%. Note that Excel does not provide probabilities to test the null (r=0) hypothesis.

Results when you use SPSS to calculate a Correlation

Correlations

		HOURS	GPA
HOURS	Pearson Correlation	1.000	**.915**
	Sig. (2-tailed)	.	.000
	N	10	10
GPA	Pearson Correlation	**.915**	1.000
	Sig. (2-tailed)	.000	.
	N	10	10

** Correlation is significant at the 0.01 level (2-tailed).

Unlike Excel, SPSS gives you the correlation as well as the significance (or p-value) associated with the correlation. Once again, they have filled in 1.000 when comparing a variable to itself. You can ignore those entries.

Summary

Linear regression and correlation are closely related, but there are some differences between them to help us distinguish which one is most appropriate for a given situation. Linear regression is used for prediction purposes and calculates an equation for the "best-fit" line which answers the question, "what is the relationship between X and Y?". Correlation analysis focuses on the strength of the relationship between X and Y or how tightly the data points are clustered around the "best-fit" line.

Chapter Ten Homework

Provide a short answer for the following questions.

1. What is R^2 (or the coefficient of determination) and how do you calculate it?

2. What is R^2 (the coefficient of determination) for a correlation of $r = 0.89$?

3. How would you put the answer to question #2 into words (i.e., what does it tell us about the variability of X and Y)?

4. What is the possible range for a correlation coefficient?

5. What two considerations must you take into account when deciding which correlation coefficient to use?

6. What assumption does Pearson's r make about the relationship between X and Y?

7. Researchers are interested in whether frequent club attendance affects hearing. Ten volunteers from the community are randomly selected; the experimenters measure their average monthly club attendance (variable X) and their hearing pain thresholds (their maximum comfortable volume in decibels, variable Y). For the data below, calculate the appropriate correlation between the two variables.

$\Sigma X=$	19
$\Sigma X^2=$	71
$\Sigma XY=$	1,782
$\Sigma Y=$	1,035
$\Sigma Y^2=$	108,293

8. Find the least squares regression line for the data in #7.

9. Correlation and regression differ in that ___________

(a) correlation is primarily concerned with the size and direction of relationships
(b) regression is primarily used for prediction
(c) both a and b are true
(d) neither a nor b are true

10. In an imperfect relationship, ___________

(a) all the points fall on the line
(b) a relationship exists, but all of the points do not fall on the line
(c) no relationship exists
(d) a relationship exists, but none of the points can fall on the line

11. The closer the points on a scatter diagram fall to the regression line, the ___________ between the scores

(a) higher the correlation
(b) lower the correlation
(c) correlation doesn't change
(d) need more information

12. The lowest degree of correlation shown below is ___________

(a) 0.75
(b) –0.33
(c) –0.25
(d) 0.15

13. Knowing nothing more than that IQ and memory scores are correlated 0.84, you could validly conclude that _______________

(a) good memory causes high IQ
(b) high IQ causes good memory
(c) neither good memory nor high IQ cause each other
(d) a third variable causes both good memory and high IQ
(e) none of the above

14. A correlation between college entrance exam grades and scholastic achievement was found to be –1.08. On the basis of this you would tell the university that .....

(a) the entrance exam is a good predictor of success
(b) they should fire that sorry statistician and hire a new one
(c) the exam is a poor predictor of success
(d) students who do best on this exam will make the worst students
(e) students at this school are underachieving

15. An experimenter believes that reaction time on a word-identification task might be related to one's score on a vocabulary test. She tested a random sample of people from a normal population. Reaction time was measured in milliseconds and the score on a vocabulary test was out of 100 points.

$\Sigma X=$	5,290
$\Sigma X^2=$	2,935,750
$\Sigma XY=$	420,070
$\Sigma Y=$	804
$\Sigma Y^2=$	65,382
$N=$	10

$S_{y/x} =$ 8.22

(a) Calculate the appropriate correlation coefficient

(b) Calculate the slope of the regression line

(c) Calculate the y-intercept of the regression line

(d) Given the reaction time of 400 ms, predict this person's score on the vocabulary test.

(e) Based on the information you have, how confident are you about your prediction?

CHOOSE THE MOST APPROPRIATE AND POWERFUL TEST:

16. A sports psychologist analyzed the batting average of subjects that hit right-handed (n=59), subjects that hit left-handed (n=41) and subjects that hit from either side (switch hitters! n = 30). The subjects were randomly selected from active major league players across the American and National Leagues. The variances are 0.18, 0.24 & 0.32 respectively.

17. Research on reaction time indicates that, on average, people press a button after a bell goes off in 1.88 seconds. A random sample of 16 subjects from a normally distributed population of airline pilots was assessed using the reaction-time test to determine if the reaction time of pilots is significantly faster than the reaction time in the general

population. Their mean reaction time was 1.61 seconds with a standard deviation of 0.07 seconds.

18. Over a 50 year period, the number of children per couple in China has been 3.88, with a standard deviation of 0.91. Recently, China has tried to decrease the number of children born per couple. A sample of 30 couples living in China was performed, and they had an average of 1.38 children with a standard deviation of 0.25. Have the new policies resulted in a significant change in the number of children per family?

19. A researcher is interested to know whether there is any difference in spatial aptitude for air-traffic controllers and artists. Thirty-five air-traffic controllers and 35 artists were randomly selected to participate in the study. Both groups were given spatial skills test in which the number of questions answered correctly were evaluated. Air-traffic controllers averaged a mean of 23 problems right with a standard deviation of 4. Artists averaged a mean of 27 problems right with a standard deviation of 6. Can the researcher conclude that air-traffic controllers and artists differ in their spatial aptitude?

20. A researcher performed a study examining the impact divorce has on anxiety. She randomly sampled 101 married individuals, 31 of which later became divorced. To measure anxiety, she recorded heartbeats per minute of the 31 divorced individuals before and after they were divorced. How should she analyze the data?

SIDEBAR 1: Calculating the y-intercept and the slope manually
Where did Excel come up with the values for the y-intercept and the slope? These can be calculated using the following formulas:

$$b_y = \frac{\sum XY - \frac{(\sum X)(\sum Y)}{n}}{\sum X^2 - \frac{(\sum X)^2}{n}}$$

$$a_y = \overline{Y} - b_y \overline{X}$$

where:
n = number of pairs of scores
ΣXY = Sum of the product of each X and Y pair, or the sum of the cross products.

Let's apply this to our previous example to illustrate calculating the y-intercept and the slope without the aid of a computer.

Step one: Calculate the intermediate numbers required for the formula

Student	Hours(X)	X^2	Grade(Y)	Y^2	X * Y
1	10	100	3.4	11.56	34
2	12	144	3.5	12.25	42
3	5	25	2.7	7.29	13.5
4	2	4	1.7	2.89	3.4
5	20	400	3.9	15.21	78
6	14	196	3.6	12.96	50.4
7	11	121	3.3	10.89	36.3
8	22	484	3.8	14.44	83.6
9	3	9	1.9	3.61	5.7
10	18	324	4	16	72
Sums	117	1807	31.8	107.1	418.9

$$\overline{Y} = \frac{31.8}{10} = 3.18$$

$$\overline{X} = \frac{117}{10} = 11.7$$

Step Two: Use the formulas to calculate the y-intercept and slope

$$b_y = \frac{418.9 - \frac{(117)(31.8)}{10}}{1807 - \frac{(117)^2}{10}} = \frac{46.84}{438.1} = 0.1069$$

$a_y = 3.18 - (0.1069)(11.7) = 1.929$

Step Three: Insert the y-intercept and slope values into the regression line equation

Y' = 0.1069(X) + 1.929

An X value of 13 would yield a Y value of 3.3.

Y' = 0.1069(13) + 1.9291 = 3.3188

Based on the information provided in this dataset, your best estimate of the GPA of a student who studied 13 hours a week would be 3.3.

SIDEBAR 2: Formula for calculating the Standard Error of the Estimate (SEE)

$$s_{y/x} = \sqrt{\frac{SS_Y - \frac{[\sum XY - \frac{(\sum X)(\sum Y)}{n}]^2}{SS_X}}{n-2}}$$

SIDEBAR 3: Formula for Pearson's r

$$r = \frac{\sum XY - \frac{(\sum X)(\sum Y)}{n}}{\sqrt{\left[\sum X^2 - \frac{(\sum X)^2}{n}\right]\left[\sum Y^2 - \frac{(\sum Y)^2}{n}\right]}} = \frac{\sum XY - \frac{(\sum X)(\sum Y)}{n}}{\sqrt{(SS_X)(SS_Y)}}$$

Note that the formula requires that you calculate the sums-of-squares for both X and Y, but the numerator contains a formula that involves the cross products of X and Y that we used in the linear regression calculations. In fact, the numerator of the Pearson's r coefficient formula is exactly the same as the numerator of the slope. No wonder the sign of the Pearson's r coefficient indicates the direction of the relationship or slope!

Calculating the Pearson's r Coefficient for the Effort and GPA Example

Step One: Calculate the intermediate numbers required for the formula

Student	Hours(X)	X^2	Grade(Y)	Y^2	X * Y
1	10	100	3.4	11.56	34
2	12	144	3.5	12.25	42
3	5	25	2.7	7.29	13.5
4	2	4	1.7	2.89	3.4
5	20	400	3.9	15.21	78
6	14	196	3.6	12.96	50.4
7	11	121	3.3	10.89	36.3
8	22	484	3.8	14.44	83.6
9	3	9	1.9	3.61	5.7
10	18	324	4	16	72
Sums	117	1807	31.8	107.1	418.9

$$SS_X = 1807 - \frac{(117)^2}{10} 438.1$$

$$SS_Y = 107.1 - \frac{(31.8)^2}{10} = 5.976$$

Step Two: Use the Pearson's r formula to calculate the coefficient

$$r = \frac{\sum XY - \frac{(\sum X)(\sum Y)}{n}}{\sqrt{(SS_X)(SS_Y)}}$$

$$r = \frac{418.9 - \frac{(117)(31.8)}{10}}{\sqrt{(438.1)(5.976)}} = \frac{46.84}{51.167232} = 0.9154296$$

Chapter 11
General Linear Model

The General Linear Model is a relatively new (last 20-30 years) approach to parametric statistics. Statisticians have been rethinking both the use and the interpretation of the common statistical tests, such as ANOVA and t-tests, and realized that all of these statistics have a common mathematical underpinning. We have been foreshadowing this common underpinning in several of the previous chapters of the book. Here we will formally describe this relationship and why it is useful to us.

Linear or Additive Equation/Model

This concept was first discussed in chapter 8 (ANOVA) where we said that each score is the result of a combined influence of the grand mean (μ), any treatment effect attributable to our independent variable (IV), and unexplained error (E).

Conceptually, this states that there are multiple potential influences on the outcome for each score, and those influences can be defined with a linear equation in which each influence is added together. We did not present this information in Chapter 7 (Two-sample Tests), but both conceptually and mathematically, ANOVA is the same as a t-test, but simply with more groups. In fact, the F statistic that is calculated in an ANOVA with only two groups is simply t^2 for an Independent t-test with the same two groups. If you accept that ANOVA is actually a t-test with more groups, then it probably won't surprise you to find out that their linear equations are the same:

ANOVA / t-test

Score = μ + IV + E

The formula for regression is really the same as the formula for ANOVA. The notation looks a bit different, but that is because each test was derived independently of the other test, and historically they have each had their own terminology and notation.

Regression
$Y = a + bX + E$

Where
X = X score
Y = Y score
a = Y intercept
b = slope of the line

$$\text{Slope (b)} = \frac{\Delta Y}{\Delta X} \text{ or } \frac{rise}{run}$$

Not only are t-tests, one-way ANOVA and linear regression related to one another, but complex ANOVA and multiple regression designs are also variations of this linear equation or the "General Linear Model".

Two-Way ANOVA

If there are two independent variables that could potentially influence each score, then you can define the effect of each independent variable, and the interactive effect of the two variables, conceptually with the linear equation:

$\text{Score} = \mu + IV_1 + IV_2 + (IV_1)(IV_2) + E$

Where
μ = Combined influence of the grand mean
IV_1 = Effect of the first independent variable
IV_2 = Effect of the second independent variable
$(IV_1)(IV_2)$ = Interaction effect of the independent variables
E = Unexplained error

Three-Way ANOVA

If you are measuring three independent variables (a three-way ANOVA), then an individual's score could be represented in the following way:

$$\text{Score} = \mu + IV_1 + IV_2 + IV_3 + (IV_1)(IV_2) + (IV_1)(IV_3) + (IV_2)(IV_3) + (IV_1)(IV_2)(IV_3) + E$$

Where
μ = Combined influence of the grand mean
IV_1 = Effect of the first independent variable
IV_2 = Effect of the second independent variable
IV_3 = Effect of the second independent variable
$(IV_1)(IV_2)$ = Interaction effect of the first two independent variables
$(IV_1)(IV_3)$ = Interaction effect of the first and third independent variables
$(IV_2)(IV_3)$ = Interaction effect of the second and third independent variables
$(IV_1)(IV_2)(IV_3)$ = Interaction effect of all three independent variables
E = Unexplained error

Repeated Measures ANOVA

Another variation of the ANOVA is the Repeated Measures ANOVA where subjects serve in more than one condition of the experiment, or subjects are matched so that they are treated (statistically) as if they are the same individual.

$$\text{Score} = \mu + IV_1 + \text{Subj} + E$$

Where
μ = Combined influence of the grand mean
IV_1 = Effect of the independent variable
Subj = Effect attributable to individual differences between subjects
E = Unexplained error

Multiple Regression

Multiple regression is like simple regression but with an additional continuous independent variable. It is the regression parallel to Two-Way ANOVA.

$Y = a + b_1X_1 + b_2X_2 + E$

Where
X_1 = X score for the first independent variable
X2 = X score for the second independent variable
Y = Y score
a = Y intercept
b_1 = slope of the first line
b_2 = slope of the second line

All of these formulas share in common similar assumptions (i.e., normal distribution) and a linear form where the effects of each variable on the score are additive. Thus, all of these equations are fundamentally the same form of linear equation or model, and they can all be solved using the same process (matrix algebra).

Application of the General Linear Model Approach

What advantages can we gain by conceptually understanding that all of these tests are based on the General Linear Model (GLM)? We can gain two very practical advantages.

1) *You can understand and interpret why there is an F-obtained value in your linear regression output.* Modern statistical software will automatically give you an F-obtained value when you conduct a regression analysis. The F value indicates how well the slope and intercept of your regression line fits your data. It is an overall analysis of how well this line fits your data, much like the ANOVA is an overall test of more than two means. Likewise, you will also often see t-obtained values on the output. These t values are like multiple comparisons were for one-way ANOVA because they are testing how well individual parts (like your slope or your

Y-intercept) fit your overall model, and you do not need to look at them unless your F-obtained (or overall test) is significant.

2) *You can mix and match the measurement scales of your independent variables.* Until this point we have kept ANOVA and regression fairly separate in our discussions, stating that ANOVA was suited to situations where you could group your independent variable into nominal categories (sex, age groups, drug dosages, etc.) and that regression was appropriate when your independent variable was continuous (or interval/ratio). However, you might come up with an experimental design with more than two independent variables where one of them is nominal and one of them is continuous. In this case, neither the classic ANOVA design nor a regression design are appropriate, yet an overall test is still desired. You might wish to compare the effects of an antidepressant medication based on age (interval scale) and dose (broken into control, low and high dose groups). Using a subject's actual age rather than grouping subjects by broad age ranges (like <20, 20-40, 40-60, >60) is more powerful statistically because you are not collapsing your high-resolution (interval/ratio) data into lower-resolution (ordinal) categories.

SPSS Example

A clinical psychologist tests the effects of three levels of an anti-depressant medication on the number of days that subjects report feeling depressed. The subject's age is included as an additional independent variable of interest as there have been some reports that treatment can be influenced by the age of the subject, and the researcher is particularly interested in any interactions between dose and age. Table 11.1 contains the raw data entered into SPSS, and the subsequent tables are abbreviated versions of the SPSS output from the analsyis. The analysis was run in SPSS version 10.0 for Windows using the following commands (see Figures 11.1 – 11.4):

Figure 11.1 Raw Data File for SPSS

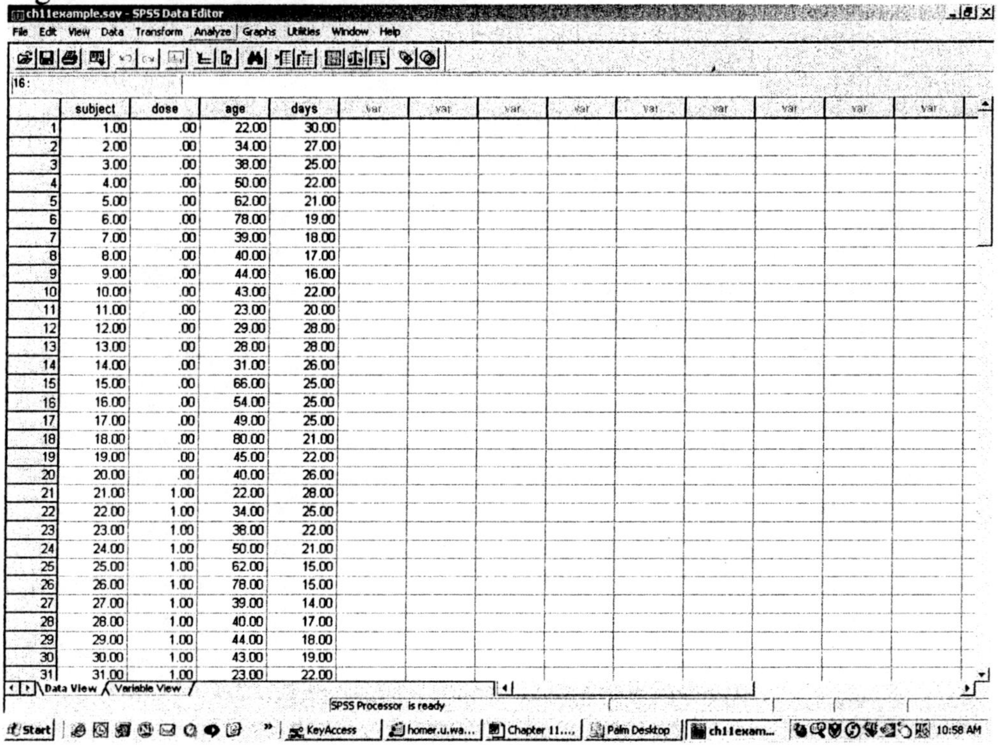

	subject	dose	age	days	var	var	var	var	var	var	var	var	var
1	1.00	.00	22.00	30.00									
2	2.00	.00	34.00	27.00									
3	3.00	.00	38.00	25.00									
4	4.00	.00	50.00	22.00									
5	5.00	.00	62.00	21.00									
6	6.00	.00	78.00	19.00									
7	7.00	.00	39.00	18.00									
8	8.00	.00	40.00	17.00									
9	9.00	.00	44.00	16.00									
10	10.00	.00	43.00	22.00									
11	11.00	.00	23.00	20.00									
12	12.00	.00	29.00	28.00									
13	13.00	.00	28.00	28.00									
14	14.00	.00	31.00	26.00									
15	15.00	.00	66.00	25.00									
16	16.00	.00	54.00	25.00									
17	17.00	.00	49.00	25.00									
18	18.00	.00	80.00	21.00									
19	19.00	.00	45.00	22.00									
20	20.00	.00	40.00	26.00									
21	21.00	1.00	22.00	28.00									
22	22.00	1.00	34.00	25.00									
23	23.00	1.00	38.00	22.00									
24	24.00	1.00	50.00	21.00									
25	25.00	1.00	62.00	15.00									
26	26.00	1.00	78.00	15.00									
27	27.00	1.00	39.00	14.00									
28	28.00	1.00	40.00	17.00									
29	29.00	1.00	44.00	18.00									
30	30.00	1.00	43.00	19.00									
31	31.00	1.00	23.00	22.00									

Figure 11.2 Choosing the General Linear Model Option in SPSS

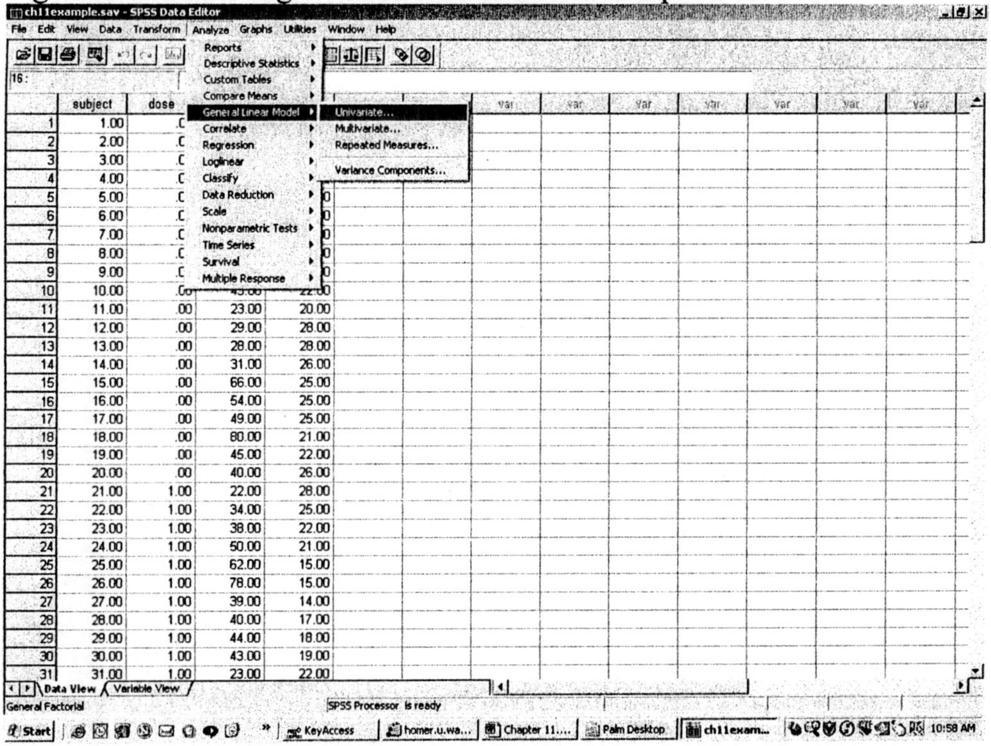

The initial steps to conduct a two-way ANOVA in SPSS were:
Select “Analyze” from the Toolbar
Select “General Linear Model”
Select “Univariate”

Figure 11.3 Selecting your Independent and Dependent Variables in SPSS

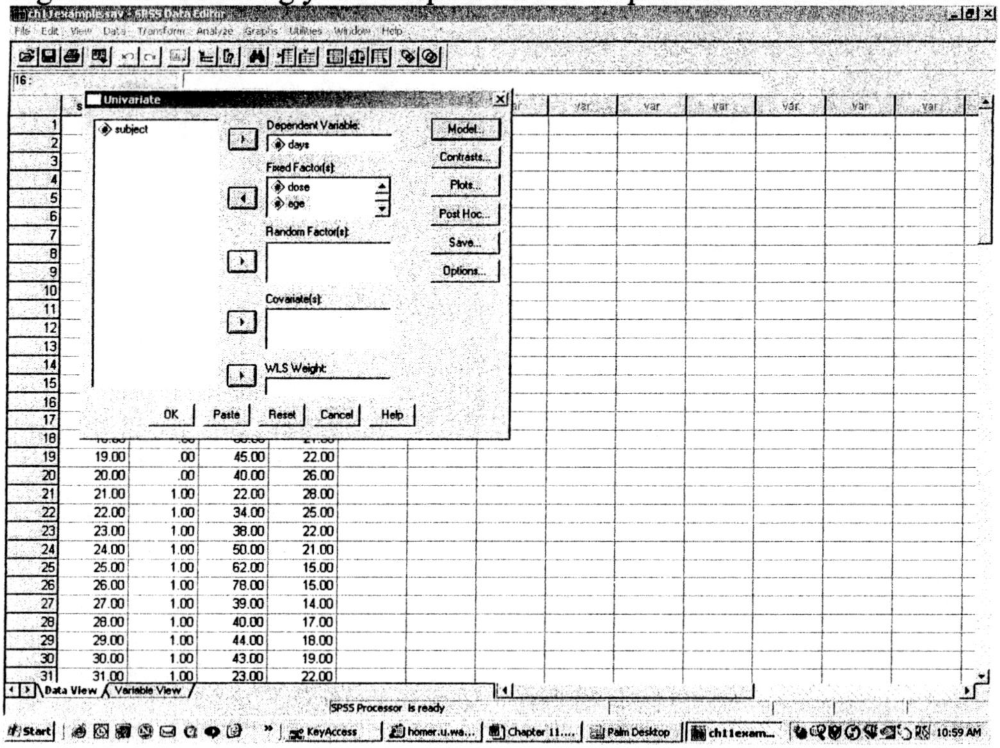

Steps in SPSS:
Select "Days" as Dependent Variable
Select "Age" and "Dose" as the Fixed Factors

Note that the "Univariate" test is referring to the fact that we are only analyzing ONE dependent variable. If there were more than one dependent variable, this would be considered a multivariate ANOVA.

Figure 11.4 Choosing the model in SPSS

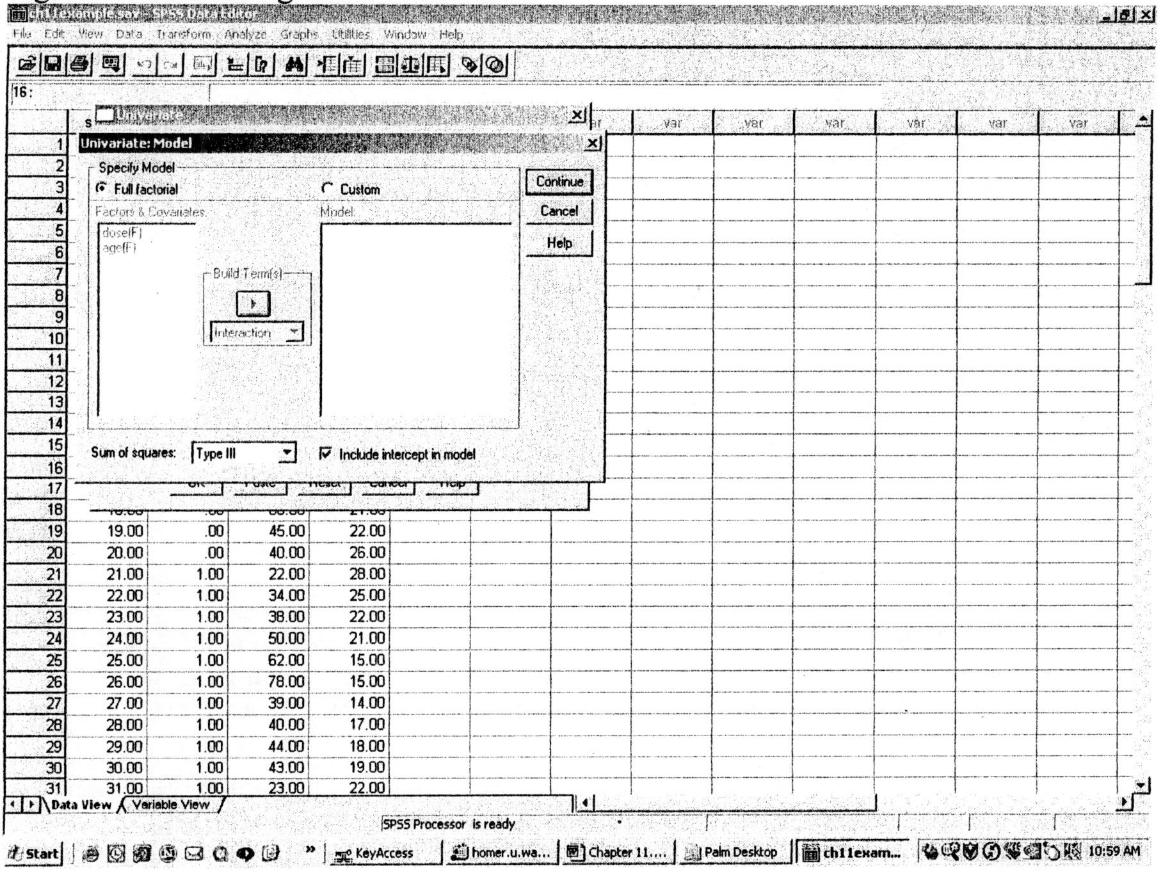

SPSS will use the default model of a "Full-factorial" design. This is what you want because it includes both of the main effects and the interaction term. You can also click the box to include a test of the y-intercept in the analysis. The output has been slightly condensed from SPSS.

Table 11.1

Tests of Between-Subjects Effects
Dependent Variable: DAYS

Source	Type III Sum of Squares	df	Mean Square	F	Sig.
Corrected Model	4005.000	70	57.214	1.599	.228
Intercept	18992.757	1	18992.757	530.853	.000
DOSE	**1165.244**	**2**	**582.622**	**16.284**	**.001**
AGE	**1063.032**	**29**	**36.656**	**1.025**	**.520**
DOSE * AGE	**862.310**	**39**	**22.111**	**.618**	**.856**
Error	**322.000**	**9**	**35.778**		
Total	26772.000	80			
Corrected Total	4327.000	79			

a R Squared = .926 (Adjusted R Squared = .347)

The main effects for DOSE and AGE are indicated on the table, and DOSE is significant (Sig. value is less than .05 alpha value), but AGE is not significant. There is no significant interaction between DOSE AND AGE (DOSE * AGE). The "Corrected Model" is an overall test of the fit of your model to the data. Note that it is not significant, most likely because both one of your main effects and your interaction are not statistically significant. These results suggest that the number of days feeling depressed varies with the dose of the antidepressant drug, but there is no relationship between age and feeling depressed and no interaction between dose and age (ie., the effect of dose is the same across all ages, or the effect of age (none) is the same at all three dose levels). We have successfully analyzed a dataset using GLM where one independent variable was continuous (Age) and would typically have been explored using traditional regression techniques, and where the second independent variable was nominal (Dose) and would have traditionally been analyzed with ANOVA.

This flexibility and the opportunity to assess the subsequent interaction between these two independent variables on different scales are the primary advantages to the GLM technique.

Analysis of Covariance (ANCOVA)

Analysis of Covariance, or ANCOVA, is a specialized form of ANOVA that is used when an investigator wishes to remove the effects of a variable that is known to influence the dependent variable, but is not the subject of the current experiment and analysis. For example, we might have added body weight to the previous example because doses can react differently in individuals of vastly different body weights, but that was not the topic of interest to the clinical psychologists. If they had added body weight into the analysis, it would be considered a covariate. They would do so in an attempt to remove that unexplained variance from the error term. This would increase their ability to detect any main effects or interactions of their independent variables. ANCOVA has been replaced by GLM. By adding an additional variable, you can achieve the desired effect of partitioning out the variance associated with it. SPSS has a separate category into which you can place those variables (see Figure 11.3). This is advantageous because it will automatically be excluded from interaction terms, in which you are not likely to be interested since body weight was not the subject of the study.

Summary

The general linear modeling approach is very powerful, allowing the simultaneous assessment of continuous and categorical independent variables. The concept of general linear modeling has been widely embraced by the research community but is still rarely included in undergraduate statistics textbooks. More importantly for this course, these concepts help us to interpret the results of statistical software packages that use the GLM terminology, which has become universal.

Chapter Eleven Homework

Provide a short answer for the following questions.

1. What are two practical advantages to using a GLM approach?
2. Why is an analysis with two independent variables and one dependent variable considered a UNIVARIATE test in SPSS?
3. What does ANCOVA stand for?
4. What is the purpose of ANCOVA?
5. How are ANOVA, Regression and t-tests similar to one another?
6. Why haven't we always known about GLM and why aren't the terms in all of the formulas in #5 all the same?

Chapter Twelve
Nonparametric Tests

Parametric tests, such as t-tests, z-tests and F-tests, require that populations are normally distributed and, in the case of independent t-test and ANOVA, that the variances are equal. Nonparametric tests have few assumptions but are much less powerful than parametric tests. Thus, you should use parametric tests whenever possible. But, when an assumption is extremely violated or the data are from the nominal or ordinal measurement scale, then you should use nonparametric tests. Despite being less powerful than parametric tests, nonparametric tests are useful in certain circumstances and are easy to calculate. We will begin with the appropriate test to use when your data are from the nominal measurement scale, and then turn to tests that use the ordinal scale. Keep in mind that these tests are appropriate either when your data are originally collected on these scales, or when you can reduce data from an interval or ratio scale down to a nominal or ordinal scale.

Chi-Square

The chi-square test is appropriate when you have nominal independent variable(s) but only group membership (frequencies) as your raw data. For example, you might want to determine whether students in this class are randomly distributed in their political party affiliations (see Table 12.1). Conceptually, this is an appropriate analysis for a chi-square test because the purpose of a chi-square analysis is to compare the *observed* frequencies to the *expected* frequencies. The *expected* frequencies are determined by assuming that group membership is random. If membership in a group differs from random, then we can say that there are statistically significant differences between the actual data and the results that one would expect due to chance (or random factors).

Table 12.1
Number of students that belong to each of the four political parties
(*Observed* frequencies)

Democrat	Green	Independent	Republican
40	20	12	40

N = 40 + 20 + 12 + 40 = 112 students in the course

Calculation of expected values

If the expected frequencies are based purely on chance, then they are easy to calculate based on how many categories you have in your analysis. In the political party example above, we know that we have four categories of political parties, and the probability of being in each category simply due to chance would be ¼ or 25%. This is the random model of calculating *expected* frequencies.

There is a second form of expected frequencies, called a "goodness-of-fit" model. In this method, you calculate expected frequencies based on some prior knowledge of what those values should be that are not necessarily random (the goodness-of-fit model). For example, you might wish to determine whether or not the political party frequencies you observe now are significantly different than they were in the past, so you want to compare today's frequencies (observed frequencies) with the frequencies from, say, 30 years ago, and use the historical percentages to calculate your *expected* frequencies. This technique is used frequently in introductory biology or genetics courses where basic Mendelian genetics places traits into nominal categories (like dominant or recessive). For example, it is known that ¾ (75%) of the normal population has the ability to curl their tongue, while ¼ (25%) typically cannot curl their tongue (Table 12.2). This is a dominant inherited trait.

Table 12.2
Expected frequencies of tongue curling based on Mendelian genetics

Can Curl Tongue	Cannot Curl Tongue
75%	25%

Clearly this trait is not expected to be inherited based on the random model, which would predict that the ability or inability to curl your tongue would have expected frequencies of 50% respectively. This is a clear case where the goodness-of-fit model of calculating *expected* frequencies is more appropriate. We will return to both the random and the goodness-of-fit models of calculating expected frequencies when we present two complete examples of chi-square tests, but first we turn to the formula for calculating a chi-square test.

Calculation of the Chi-square (χ^2) Test

f_e = expected frequency with random sampling from the null hypothesis population

f_o = observed frequency in the sample

k = number of cells (or categories)

$$\chi^2 \text{obtained} = \sum_{cell=1}^{cell=k} \left[\frac{(f_o - f_e)^2}{f_e} \right]$$

The formula requires that you perform the calculations indicated in the brackets for each cell (or category) and then sum the numbers you calculated for each cell (or category) to produce the chi-square (χ^2) obtained value. Once that is done, you must compare the obtained value to a critical value in the same manner than you have done for the parametric tests.

If χ^2 obtained $\geq \chi^2$ critical, then you reject the null hypothesis. If the obtained value is not greater or equal to the critical value, then you fail-to-reject the null hypothesis. A table of critical values for χ^2 can be found at the back of the book (Table F). Critical values for χ^2 are based on the degrees of freedom (k – 1) and the alpha level. Evaluations of χ^2 obtained are always nondirectional.

Complete Example

A statistics professor asked the students in her course to indicate their political party affiliation in an anonymous survey. She found that 40 of her students considered themselves Democrats, 20 were Green Party members, 12 supported the Independents and 40 were Republicans out of her 112 total students in the course. Is party affiliation distributed randomly across the four parties represented by her class?

Null hypothesis (H_o): Political party affiliations are distributed randomly.

Alternative hypothesis (H_A): Political party affiliations are not randomly distributed.

Step One: Determine the *expected* frequency (f_e)

Based on the story problem, it is clear that we are supposed to compare the actual party affiliations of this class with a random model. The f_e for this example would be ¼ or 25% (one divided by the number of cells) because there are four possible cells and if the results were random then one would expect that each cell (or category) is equally likely.

Cell 1	Cell 2	Cell 3	Cell 4
25%	25%	25%	25%

In order to convert these percentages to a frequency, we need to multiply the percentage by the total N.

$f_e = 112 \times 0.25 = 28$

Thus, we have the same expected frequency for each cell.

Step Two: Calculate the χ^2 obtained

	Democrat	Green	Independent	Republican
f_o	40	20	12	40
f_e	28	28	28	28

$$\chi^2 \text{obtained} = \sum_{cell=1}^{cell=k} \left[\frac{(f_o - f_e)^2}{f_e} \right]$$

$$\chi^2\text{obtained} = \left[\frac{(40-28)^2}{28}\right] + \left[\frac{(20-28)^2}{28}\right] + \left[\frac{(12-28)^2}{28}\right] + \left[\frac{(40-28)^2}{28}\right]$$

$$\chi^2\text{obtained} = \left[\frac{(12)^2}{28}\right] + \left[\frac{(-8)^2}{28}\right] + \left[\frac{(-16)^2}{28}\right] + \left[\frac{(12)^2}{28}\right]$$

$$\chi^2\text{obtained} = \left[\frac{144}{28}\right] + \left[\frac{64}{28}\right] + \left[\frac{256}{28}\right] + \left[\frac{144}{28}\right]$$

$$\chi^2\text{obtained} = 9.142857 + 5.142857 + 5.142857 + 2.285714 = 21.71429$$

Step Three: Evaluating the probability of obtaining this χ^2 value due to chance

To evaluate your χ^2-obtained value, you must use the χ^2 distribution (Table F), and to do so, you need an alpha level (0.05) and degrees of freedom (k-1). Thus, the df for this problem are 4-1 = 3. Compare the χ^2-critical value with your χ^2-obtained value, and reject the null hypothesis when your χ^2-obtained value is equal to or greater than your χ^2-critical value. When $\alpha = 0.05$ and the degrees of freedom are equal to 3, you should use 7.815 as your χ^2-critical value. $21.71 \geq 7.815$, so we *reject* the null hypothesis.

How should we interpret these data in light of the randomness of party affiliation for the students in this hypothetical course?

These results suggest that you would expect to obtain these frequencies in party affiliation less than 5% of the time, assuming party affiliation was random. Thus, it is not very likely that these frequencies all come from the random null hypothesis population. Of course, as always, there is some finite probability that you could get frequencies that deviate from the expected frequencies this much simply due to chance, but that probability is less than 5%.

Results of Chi-square test using SPSS

NPar Tests
Chi-Square Test
PARTY

	Observed N	Expected N	Residual
1.00	**40**	**28.0**	12.0
2.00	**20**	**28.0**	-8.0
3.00	**12**	**28.0**	-16.0
4.00	**40**	**28.0**	12.0
Total	**112**		

Test Statistics

	PARTY
Chi-Square	**21.714**
df	**3**
Asymp. Sig.	.000

a. 0 cells (.0%) have expected frequencies less than 5. The minimum expected cell frequency is 28.0.

We have highlighted the numbers that are comparable to our manual calculations. The exact probability (SPSS calls it "Sig.") is less than our alpha level of 0.05, thus we must assume that these results would be *unlikely* simply due to chance. SPSS does not give you a critical value to compare with your obtained value, but it is not necessary since SPSS does give you the exact probability of obtaining these results or results more extreme than these simply due to chance.

How would these results be reported in a scientific journal article?

If the journal required American Psychological Association (APA) format, the results would be reported in a format something like this:

> Political party affiliations differ significantly from random, χ^2 (3, N=112) = 21.71, $p < .01$.

This formal sentence includes the independent variable (party affiliation), as well as a statement about statistical significance, the symbol of the test (χ^2), the degrees of freedom (3), the statistical value (21.71), and the estimated probability of obtaining this result simply due to chance (<.01).

2

Second Example (f_e calculated with the Goodness-of-fit Model)

The same statistics professor now asks the students in her course to indicate whether or not they can curl their tongues. She finds that 80 of her students can curl their tongues, while 32 of her 112 total students cannot curl their tongues even with practice. Does this sample of students fit the Mendelian pattern that a single gene dominant trait is exhibited by 75% of individuals while the recessive trait is exhibited by 25% of all individuals? Remember that we only have to show that there is no significant difference between our sample and the goodness-of-fit model.

Null hypothesis (H_o): Tongue curling in this sample fits the expected model of 75% and 25%.

Alternative hypothesis (H_A): This sample does not come from the normal population where 75% of all individuals can curl their tongues and 25% of all individuals cannot curl their tongues.

Step One: Determine the *expected* frequency (f_e)
Based on the story problem, it is clear that we are supposed to compare the actual frequencies of tongue curling for this class with a goodness-of-fit model. The f_e for this example would be determined by converting each percentage to a frequency. We do this by multiplying each percentage by the total N.

$f_e = 112 \times 0.75 = 84$
$f_e = 112 \times 0.25 = 28$

Thus, we have different expected frequencies for each cell.

Step Two: Calculate the χ^2 obtained

	Curl	No Curl
f_o	80	32
f_e	84	28

$$\chi^2\text{obtained} = \sum_{cell=1}^{cell=k} \left[\frac{(f_o - f_e)^2}{f_e} \right]$$

$$\chi^2\text{obtained} = \left[\frac{(80-84)^2}{84} \right] + \left[\frac{(32-28)^2}{28} \right]$$

$$\chi^2\text{obtained} = \left[\frac{(-4)^2}{84} \right] + \left[\frac{(4)^2}{28} \right]$$

$$\chi^2\text{obtained} = \left[\frac{16}{84} \right] + \left[\frac{16}{28} \right]$$

$$\chi^2\text{obtained} = 0.190476 + 0.571429 = 0.761905$$

Step Three: Evaluating the probability of obtaining this χ^2 value due to chance

To evaluate your χ^2-obtained value, you must use the χ^2 distribution (Table F), and to do so, you need an alpha level (0.05) and degrees of freedom (k-1). Thus, the df for this problem are 2-1 = 1. Compare the χ^2-critical value with your χ^2-obtained value, and reject the null hypothesis when your χ^2-obtained value is equal to or greater than your χ^2-critical value. When $\alpha = 0.05$ and the degrees of freedom are equal to 1, you should use 3.841 as your χ^2-critical value.

0.761905 < 3.841, so we *fail-to-reject* the null hypothesis.

How should we interpret these data?

These results suggest that you would expect to obtain these frequencies in tongue curling more than 5% of the time. Therefore, it is likely that these frequencies all come from the null hypothesis population where 75% of all individuals can curl their tongues and 25% cannot curl their tongues. Our sample appears to come from a population that fits the Mendelian characteristics of a single gene trait that is dominant.

How would these results be reported in a scientific journal article?

If the journal required American Psychological Association (APA) format, the results would be reported in a format something like this:

> Tongue curling in this sample is not significantly different than the expected model of 75% and 25%, χ^2 (1, *N*=112) =.76, $p >$ .05.

This formal sentence includes the independent variable (ability to curl or not to curl your tongue), as well as a statement about statistical significance, the symbol of the test (χ^2), the degrees of freedom (1), the statistical value (.76), and the estimated probability of obtaining this result simply due to chance (<.01).

Test of Independence Between Two Variables: Contingency Table Analysis

The contingency table analysis is a variation of the chi-square test that allows you to test the independence between two variables. Thus, you can ask whether two categorical variables are independent from one another, or whether they are related to one another. In order to perform this analysis, the variables must be classified into mutually exclusive categories. As we showed in the earlier chi-square discussion, the cells contain the frequencies. What is a mutually exclusive category? Mutually exclusive categories are designed so that being a member of one category prohibits your inclusion in the other category. For example, consider biological sex as a mutually exclusive category. Being male prohibits one from being in the other category (female). If one of the independent variables is called Factor A and the second one is Factor B, then we can state the null and alternative hypotheses in the following way:

Null hypothesis (H_o): Factor A is not related to Factor B.

Alternative hypothesis (H_A): Factor A and Factor B are related.

Performing a Contingency Table Analysis

The equation for the contingency table analysis is the same one that we used for the Chi-square, where you calculate a value for each cell that is determined by the observed and the expected frequencies. However, the f_e and the df are calculated differently for this analysis. The expected frequencies are slightly more challenging to calculate because you must do it for each cell, and the value is influenced by the subtotals in the rows and columns. Specifically, the f_e for each cell is calculated in the following way:

$$\frac{\text{(Row marginal for that cell)(Column marginal for that cell)}}{\text{Total N}}$$

We'll turn to an example to demonstrate the calculations and the marginal values.

Example

Hallmark Card Company hires a statistician to determine whether or not males and females differ in their preference for a "mushy" Valentine's Day card. They obtain the following results from their survey of 400 adults.

	Love 'em	Hate 'em	Don't Care	Marginals
Males	100	50	50	
Females	150	25	25	

Marginals				

Null hypothesis (H_o): Preference for "mushy" Valentine's Day cards is randomly distributed between males and females.

Alternative hypothesis (H_A): Preference for "mushy" Valentine's Day cards is NOT randomly distributed between males and females.

Step One: Determine the *expected* frequency (f_e)

Based on the story problem, it is clear that we are supposed to compare the actual frequencies of preferences for mushy Valentine's Day cards with the expected frequencies based on a random model. The expected frequencies will be based on row and column marginals, and there may be different expected frequencies for each cell.

Determine each row and column marginal by adding up the frequencies in the respective row or column.

	Love 'em	Hate 'em	Don't Care	Marginals
Males	100	50	50	200
Females	150	25	25	200
Marginals	250	75	75	400

Note that cells are numbered from left to right (marginals do not count)

	Love 'em	Hate 'em	Don't Care	Marginals
Males	Cell 1	Cell 2	Cell 3	200
Females	Cell 4	Cell 5	Cell 6	200
Marginals	250	75	75	400

Calculate Frequency Expected Values

For each cell fe = $\frac{\text{(its row marginal)(its column marginal)}}{\text{Total N}}$

$$\text{Cell 1 fe} = \frac{(200)(250)}{400} = 125$$

$$\text{Cell 2 fe} = \frac{(200)(75)}{400} = 37.5$$

$$\text{Cell 3 fe} = \frac{(200)(75)}{400} = 37.5$$

$$\text{Cell 4 fe} = \frac{(200)(250)}{400} = 125$$

$$\text{Cell 5 fe} = \frac{(200)(75)}{400} = 37.5$$

$$\text{Cell 6 fe} = \frac{(200)(75)}{400} = 37.5$$

Step Two: Calculate the χ^2 obtained

f_e values for each cell are indicated in the parentheses

	Love 'em	Hate 'em	Don't Care	Marginals
Males	100 (125)	50 (37.5)	50 (37.5)	200
Females	150 (125)	25 (37.5)	25 (37.5)	200
Marginals	250	75	75	400

$$\chi^2\text{obtained} = \sum_{cell=1}^{cell=k}\left[\frac{(f_o - f_e)^2}{f_e}\right]$$

χ^2obtained =

$$\left[\frac{(100-125)^2}{125}\right]+\left[\frac{(50-37.5)^2}{37.5}\right]+\left[\frac{(50-37.5)^2}{37.5}\right]+\left[\frac{(150-125)^2}{125}\right]+$$

$$\left[\frac{(25-37.5)^2}{37.5}\right]+\left[\frac{(25-37.5)^2}{37.5}\right]=$$

$$\chi^2\text{obtained}=\left[\frac{(-25)^2}{125}\right]+\left[\frac{(12.5)^2}{37.5}\right]+\left[\frac{(12.5)^2}{37.5}\right]+\left[\frac{(25)^2}{125}\right]+$$

$$\left[\frac{(-12.5)^2}{37.5}\right]+\left[\frac{(-12.5)^2}{37.5}\right]=$$

$$\chi^2\text{obtained}=\left[\frac{625}{125}\right]+\left[\frac{156.25}{37.5}\right]+\left[\frac{156.25}{37.5}\right]+\left[\frac{625}{125}\right]+\left[\frac{156.25}{37.5}\right]+\left[\frac{156.25}{37.5}\right]=$$

$$\chi^2\text{obtained} = 5 + 4.1666 + 4.1666 + 5 + 4.1666 + 4.1666 = 26.6666$$

Step Three: Evaluating the probability of obtaining this χ^2 value due to chance

To evaluate your χ^2-obtained value, you must use the χ^2 distribution (Table F), and to do so, you need an alpha level (0.05) and degrees of freedom [df = (r-1)(c-1)], where r = the number of rows and c = the number of columns. Thus, the df for this problem are (2-1)(3-1) = 2. Compare the χ^2-critical value with your χ^2-obtained value, and reject the null hypothesis when your χ^2-obtained value is equal to or greater than your χ^2-critical value. When $\alpha = 0.05$ and the degrees of freedom are equal to 2, you should use 5.991 as your χ^2-critical value. 26.67 > 5.991, so we *reject* the null hypothesis.

How should we interpret these data?

These results suggest that you would expect to obtain these

frequencies in valentine card preference less than 5% of the time. Thus, it is unlikely that these frequencies come from the null hypothesis population where card preferences are distributed randomly by sex. It appears that sex and preference for mushy Valentine's Day cards are related. In fact, it appears that women tend to prefer mushy Valentines (!).

How would these results be reported in a scientific journal article?

If the journal required American Psychological Association (APA) format, the results would be reported in a format something like this:

> There was a significant relationship between sex and a preference for mushy Valentine's Day cards, χ^2 (2, *N*=400) =26.67, $p < .01$.

This formal sentence includes both independent variables (card preference as well as sex), a statement about statistical significance, the symbol of the test (χ^2), the degrees of freedom (2), the statistical value (26.67), and the estimated probability of obtaining this result simply due to chance (<.01).

Assumptions for the Chi-Square & Contingency Table Tests (χ^2)

There are two assumptions for the chi-square test:

1. Each subject has only one entry and the categories are mutually exclusive. This ensures that there is independence between your observations.
2. The expected frequency in each cell must be *at least five* when the rows or columns are greater than two.

The chi-square test is frequently used with nominal data but you can convert other scales into mutually exclusive categories and frequencies to employ the chi-square test.

Nonparametric Tests to Perform when the Parametric Test Cannot be Performed Due to Violation of the Assumptions

There are alternative tests that can be conducted when one or more of the assumptions of a parametric test are not met. These alternatives do not require a normal distribution or homogeneity of variance or interval/ratio data (although they STILL require an assumption of random sampling). These tests are typically performed when the assumptions are not met or when the data collected are only on an ordinal scale. However, they can be conducted on data that are originally collected on interval/ratio scale, because they convert the data to an ordinal scale as part of the calculations.

Table 12.3
Non-parametric equivalents of popular parametric tests

Parametric Test	Equivalent Nonparametric Test
Correlated (paired) t-test	Wilcoxon
Independent t-test	Mann-Whitney U
One-Way ANOVA	Kruskal-Wallis

Wilcoxon Signed Rank Test (T)

The Wilcoxon Signed Rank Test (T) replaces the correlated (or paired) t-test. This test considers both the magnitude and the direction of the effect, but it is not as powerful as the t-test. This test is appropriate when you have a nominal independent variable with two groups (i.e., before versus after), but your dependent variable is *at least* on an ordinal scale.

Calculation Steps

a. Calculate the differences between the two conditions (e.g. before versus after). This is your difference score; the Wilcoxon test is similar to the paired t-test at this point.
b. Rank the absolute values of the difference scores from the smallest to the largest. If there are ties on the ranked scores, then you should take an average rank, but do not re-use ranks. This will be clearer when you see an example.
c. Assign the sign of the original difference score to the rank (put the signs back on). So if your difference score was negative, you should add that negative back on to the ranked score so that you have a negative rank.
d. Sum all of the positive ranks together and then separately sum the negative ranks. Calculate the absolute value of the summed ranks for each group.
e. The smaller of the summed ranks becomes the T obtained value.
f. Evaluate with Wilcoxon table:
    a. NOTE: reject if $T_{obt} \leq T_{crit}$

(1) **Complete Example**

A new program is developed to enrich the kindergarten experience of children in preparation for first grade. Compare the test scores of nine children before and after the program is initiated. Note that a score of 10 indicates that the child is performing at the grade level. Compare the pre-test and post-test scores for these children.

Pupil	Pre-test	Post-test
A	9	16
B	7	12
C	14	13
D	12	10
E	8	14
F	7	11
G	12	15
H	8	10
I	12	14

Step One: Calculate the difference scores

Note that since we are subtracting from left to right, increased scores on the post-test result in negative difference scores (suggesting improvement in the scores). You could also subtract in the opposite direction. As long as you are consistent, it won't affect the result.

Pupil	Pre-test	Post-test	Difference
A	9	16	-7
B	7	12	-5
C	14	13	1
D	12	10	2
E	8	14	-6
F	7	11	-4
G	12	15	-3
H	8	10	-2
I	12	14	-2

Step Two: Rank the absolute value of the difference score

Organize the raw difference scores from smallest to largest and assign the *absolute value* of each score a rank from left to right in the following way (see note on tied ranks):

Raw Scores Only

Raw Scores	1	2	-2	-2	-3	-4	-5	-6	-7

Raw Scores with Ranks Below them

Raw Scores	1	2	-2	-2	-3	-4	-5	-6	-7
		(2)	(3)	(4)					
Rank	1	3	3	3	5	6	7	8	9

Why did we use a rank of "3" three times? If you take the absolute value of -2, then you find that you have THREE scores with an absolute value of 2. The way to handle this is to take the average of the ranks that each of the tied scores would have if they were not tied. In this case, each of these 2's hold the ranks of 2, 3 and 4. If you average these ranks $\frac{(2+3+4)}{3}$, then the average rank is 3. We use 3 for each of these, but we cannot "re-use" the ranks (2-3-4) again because these scores have already used these ranks. Instead, 5 should be the next rank.

Step Three: Assign the ranks the original sign of the difference score

Raw Scores	1	2	-2	-2	-3	-4	-5	-6	-7
Rank	1	3	-3	-3	-5	-6	-7	-8	-9

Step Four: Sum the *ranks* for the positive and negative ranks separately

Sum ranks for positive ranks

$1 + 3 = 4$

Sum ranks for negative ranks

$|-3| + |-3| + |-5| + |-6| + |-7| + |-8| + |-9| = 41$

Step Five: Determine the T obtained value

Since the smaller of the two sums is the T obtained value, in this case $T_{obt} = 4$.

Step Six: Evaluate the T obtained value

Use Table G to evaluate your T obtained value. The critical value of T is based on your paired sample size (n = 9) and your alpha level ($\alpha = 0.05$, 2-tailed). Based on this, the T critical value for this problem is 5. Note that the decision rule is that you can only reject the null hypothesis when Tobt $\leq$ Tcrit. This is backwards compared to previous tests. In this example, $4 < 5$ so we should *reject* the null hypothesis. Let's look at the SPSS output for these data that we have analyzed by hand.

Results if you use SPSS to calculate the Wilcoxon
NPar Tests

Wilcoxon Signed Ranks Test

Ranks

		N	Mean Rank	Sum of Ranks
POSTTEST PRETEST	Negative Ranks	2	2.00	**4.00**
	Positive Ranks	7	5.86	**41.00**
	Ties	0		
	Total	9		

a POSTTEST < PRETEST
b POSTTEST > PRETEST
c PRETEST = POSTTEST

Test Statistics

	POSTTEST - PRETEST
Z	-2.199
Asymp. Sig. (2-tailed)	**.028**

a Based on negative ranks.
b Wilcoxon Signed Ranks Test

We have placed the numbers that are comparable to our manual calculations in bold. The exact probability (SPSS calls it "Sig.") is less than our alpha level of 0.05, thus we must assume that these results would be *unlikely* simply due to chance. SPSS does not give you a critical value to compare with your obtained value, but it is not necessary since SPSS does give you the exact probability of obtaining these results or results more extreme than these simply due to chance.

How would these results be reported in a scientific journal article?

If the journal required American Psychological Association (APA) format, the results would be reported in a format something like this:

> There was a significant difference between the pre-and post-test scores of children after experiencing the enrichment program, $T(N=9) = 5.00, p = .03$.

This formal sentence includes the dependent variable (test scores), the independent variable (before versus after), as well as a statement about statistical significance, the symbol of the test (T), the sample size (9), the statistical value (5), and the estimated probability of obtaining this result simply due to chance (.03).

Assumptions for the Wilcoxon Test

1. Raw scores must be at least ordinal, but can be interval/ratio.
2. Must have random sampling.

Mann-Whitney U test (U)

The Mann-Whitney U test is appropriate when you have an independent (or between groups) design with two groups (e.g., experimental versus control groups). This test is designed to evaluate the separation between the groups. A larger separation suggests that the populations are unlikely to be the same, while a smaller separation suggests that both groups come from the same underlying null-hypothesis population. This test replaces the independent t-test when the original data collection is ordinal or when the assumptions of the independent t-test are not met.

Calculation Steps

a. Place scores in rank order and rank scores from both samples regardless of experimental condition. Handle tie scores by assigning the average rank as you did with the Wilcoxon test.

b. Sum the ranks separately by experimental group (R_1 and R_2).

c. Calculate two $U_{obtained}$ values by using the following equations:

$$\text{U obtained} = n_1 n_2 + \frac{n_1(n_1+1)}{2} - R_1$$

$$\text{U obtained} = n_1 n_2 + \frac{n_2(n_2+1)}{2} - R_2$$

d. The smaller value becomes the true U obtained value while the larger value is referred to as the U ' (prime) obtained value.

e. Evaluate the U obtained value using the Mann-Whitney table (Table H) and reject the null hypothesis if $U_{obt} \leq U_{crit.}$

(1) **Complete Example #1**

Does prenatal exposure to alcohol reduce SAT scores? Evaluate the SAT scores of one group of teenagers that were exposed to alcohol during prenatal development and another group of infants that were not exposed to alcohol.

Control Group vs. Experimental Group (prenatal exposure to alcohol)

Control (CONT)	Experimental (EXP)
610	340
600	410
585	365
625	315

$n_1 = 4$
$n_2 = 4$

Step One: Place scores in rank order and rank scores from both samples regardless of experimental condition

Group	Exp.	Exp.	Exp.	Exp.	Cont.	Cont.	Cont.	Cont.
RawScores	315	340	365	410	585	600	610	625
Rank	1	2	3	4	5	6	7	8

Step Two: Sum the ranks separately by experimental group (R_1 and R_2)

R_1 (Experimental Group) = 1 + 2 + 3 + 4 = 10
R_2 (Control Group) = 5 + 6 + 7 + 8 = 26

Step Three: Calculate two $U_{obtained}$ values by using the equations

$$\text{U obtained} = n_1 n_2 + \frac{n_1(n_1+1)}{2} - R_1 = 4(4) + \frac{4(4+1)}{2} - 10$$

$$\text{U obtained} = n_1 n_2 + \frac{n_2(n_2+1)}{2} - R_2 = 4(4) + \frac{4(4+1)}{2} - 26$$

Remember that the smaller value becomes the true U obtained value while the larger value is referred to as the U ‘ (prime) obtained value:

U obtained = 16 +10 – 10 = 16,
so U’ obtained = 16

U obtained = 16 + 10 – 26 = 0,
so U obtained = 0

Step Four: Check your work!

Uobt + U'obt should = $(n_1)(n_2)$

0 + 16 = 4(4) or 16 = 16

Since the two sides of the equation match, you know that you've done your work correctly.

Step Five: Evaluate the U obtained value using the Mann-Whitney table

In order to look up the U critical value, you need to know the sample sizes for each group, your α level and whether or not you are making a one-tailed or a two-tailed test of the hypothesis. Because you are asking whether prenatal exposure to alcohol *reduces* SAT scores, you are conducting a one-tailed test. Under these conditions, your U critical value is 1. You can also get the U' critical value, which is 15. We'll show you the decision rules for evaluating either the U obtained or the U ' obtained because some students prefer to use one of these rules over the other, but you should get the same conclusion regardless of which number you evaluate.

Decision Rule: If U obtained $\leq$ U critical, then reject the null hypothesis
Decision Rule: If U'obtained $\geq$ U'critical, then reject the null hypothesis.

If we use the U obtained and U obtained critical values, then $0 \leq 1$, so we can reject the null hypothesis. If we use the U ' obtained and U ' critical values, then $16 \geq 15$, so we can reject the null hypothesis. We get the same conclusion (reject the null hypothesis) regardless of which value we choose to evaluate.

Results if you use SPSS to calculate the Mann-Whitney U test
NPar Tests

Mann-Whitney Test

Ranks

	GROUP	N	Mean Rank	Sum of Ranks
SCORE	1.00	**4**	6.50	**26.00**
	2.00	**4**	2.50	**10.00**
	Total	8		

Test Statistics

	SCORE
Mann-Whitney U	**.000**
Wilcoxon W	10.000
Z	-2.309
Asymp. Sig. (2-tailed)	.021
Exact Sig. [2*(1-tailed Sig.)]	**.029**

a Not corrected for ties.
b Grouping Variable: GROUP

The first SPSS table gives you the R_1 and R_2 sums that you calculated by hand as well as the sample size in each group, while the second SPSS table indicates the $U_{obtained}$ value (0) and that the probability of obtaining this value if chance alone is operating is only 0.029.

How would these results be reported in a scientific journal article?

If the journal required American Psychological Association (APA) format, the results would be reported in a format something like this:

> There was a significant difference between the SAT scores of one group of teenagers that were exposed to alcohol during prenatal development and another group that was not exposed to alcohol, U $(N=8) = 0.00, p = .03$.

This formal sentence includes the dependent variable (SAT scores), the independent variable (alcohol exposure versus control), as well as a statement about statistical significance, the symbol of the test (U), the sample size (8), the statistical value (0), and the estimated probability of obtaining this result simply due to chance (.03).

Assumptions of the Mann-Whitney U-test

a. The dependent variable must be measured in at least the ordinal scale, but interval or ratio data can be "collapsed" into ordinal data via the ranking.
b. It is only appropriate when you have an independent, or between-groups design and two groups/conditions.
c. Random sampling is required.

Complete Example #2

A marketing researcher asks whether the desire to purchase Coca Cola after viewing a commercial varies based on the gender of the viewer. He asks the participants of the study to rate their preference for Coca Cola after viewing a commercial about Coca Cola and compares the ratings of the males and females. Because of the specific content of the commercial, he believes that it will have more appeal for females than it will have for males. He predicts that males will be less influenced by the commercial than females, and will give lower ratings compared to females.

Ratings of Coca Cola by Males and Females after Viewing a Commercial

Males (M)	Females (F)
5	10
6	10
4	8
7	5
5	8
	9
	9

$n_1 = 5$ $n_2 = 7$

Step One: Place scores in rank order and rank scores from both samples regardless of experimental condition

<u>Ties on ranks</u> - If two numbers are tied:

1. Get an average rank
2. Don't re-use ranks

Group	M	M	M	F	M	M	F	F	F	F	F	F
RawScores	4	5	5	5	6	7	8	8	9	9	10	10
		(2)	(3)	(4)			(7)	(8)	(9)	(10)	(11)	(12)
Rank	1	3	3	3	5	6	7.5	7.5	9.5	9.5	11.5	11.5

Average ranks used to calculate assigned rank

$$\frac{2+3+4}{3} = \mathbf{3}$$

$$\frac{7+8}{2} = \mathbf{7.5}$$

$$\frac{9+10}{2} = \mathbf{9.5}$$

$$\frac{11+12}{2} = \mathbf{11.5}$$

Step Two: Sum the ranks separately by experimental group (R_1 and R_2)

R_1 (Males) = 1 + 3 + 3 + 5 + 6 = 18

R_2 (Females) = 3 + 7.5 + 7.5 + 9.5 + 9.5 +11.5 + 11.5 = 60

Step Three: Calculate two $U_{obtained}$ values by using the equations

$$\text{U obtained} = n_1 n_2 + \frac{n_1(n_1+1)}{2} - R_1 \quad = 5(7) + \frac{5(5+1)}{2} - 18$$

$$\text{U obtained} = n_1 n_2 + \frac{n_2(n_2+1)}{2} - R_2 \quad = 5(7) + \frac{7(7+1)}{2} - 60$$

Remember that the smaller value becomes the true U obtained value while the larger value is referred to as the U ' (prime) obtained value:

U obtained = 35 + 15 – 18 = 32,
so U' obtained = 32
U obtained = 35 + 28 – 60 = 3,
so U obtained = 3

Step Four: Check your work!
Uobt + U'obt should = $(n_1)(n_2)$

32 + 3 = 5(7) or 35 = 35

Since the two sides of the equation match, you know that your work is done correctly.

Step Five: Evaluate the U obtained value using the Mann-Whitney table (Table H)

Because you are asking whether males are less affected by the commercial than females, you are conducting a one-tailed test. Under these conditions, your U critical value is 6. You can also get the U' critical value, which is 29.

Decision Rule: If U obtained $\leq$ U critical, then reject the null hypothesis
Decision Rule: If U'obtained $\geq$ U'critical, then reject the null hypothesis.

If we use the U obtained and U obtained critical values, then $3 \leq 6$ so we can reject the null hypothesis. If we use the U ' obtained and U ' critical values, then $32 \geq 29$, so we can reject the null hypothesis. We get the same conclusion (reject the null hypothesis) regardless of which value we choose to evaluate.

How would these results be reported in a scientific journal article?

If the journal required American Psychological Association (APA) format, the results would be reported in a format something like this:

> Males were significantly less influenced by the commercials than females, and gave lower ratings compared to females, U (N=12) = 3.00, p< .05.

This formal sentence includes the dependent variable (ranking), the independent variable (sex), as well as a statement about statistical significance, the symbol of the test (U), the sample size (12), the statistical value (3.00), and the estimated probability of obtaining this result simply due to chance (<.05).

Kruskal Wallis test (H)

The Kruskal-Wallis test replaces the one-way (independent) ANOVA when either or both of the assumptions are broken. The Kruskal-Wallis does not require that the distribution be normally distributed or that the variances between the groups are homogeneous.

Calculation Steps

a. Rank all scores from the smallest to the largest.
    i. Disregard groups initially when ranking.
    ii. Handle tied ranks as you did with the previous non-parametric tests.
b. Sum ranks for each group (R_i).
c. Calculate the $H_{obtained}$ value using the formula:

$$H_{obt} = \left[\frac{12}{N(N+1)}\right]\left[\sum\frac{(R_i)^2}{n_i}\right] - 3(N+1)$$

d. Evaluate using Chi square table, df = k-1, where k equals the number of groups in the experiment.
    i. NOTE: The decision rule returns to the typical pattern where you reject the null hypothesis if and only if, $H_{obt} \geq H_{crit.}$

(1) Complete Example

A college instructor asks whether or not the performance of her students varies with the time of day the class is held. She compares the grade point averages (GPA) of the same course topic held during three different time periods to determine if there is a difference in performance based on time of day alone.

GPA scores for one course held at three different time periods

Early am	Early pm	Late pm
2.5	2.8	3.8
3.1	3.2	3.2
2.2	3.0	4.0
2.6	2.7	3.7
2.2	2.3	3.7

Step one: Rank scores regardless of group/condition and handle ties appropriately.

Raw scores

2.2	2.2	2.3	2.5	2.6	2.7	2.8	3.0	3.1	3.2	3.2	3.7	3.7	3.8	4.0
(1)	(2)								(10)	(11)	(12)	(13)		

Rank

1.5	1.5	3	4	5	6	7	8	9	10.5	10.5	12.5	12.5	14	15

Average of ranks used to calculate absolute ranks.

$$\frac{1+2}{2} = \mathbf{1.5}$$

$$\frac{10+11}{2} = \mathbf{10.5}$$

$$\frac{12+13}{2} = \mathbf{12.5}$$

Step Two: Sum the ranks separately by experimental group

Early am	(R_1)	Early pm	(R_2)	Late pm	(R_3)
2.5	4	2.8	7	3.8	14
3.1	9	3.2	10.5	3.2	10.5
2.2	1.5	3.0	8	4.0	15
2.6	5	2.7	6	3.7	12.5
2.2	1.5	2.3	3	3.7	12.5

Sum ranks for each group (R_i)

$\Sigma R_1 = 4 + 9 + 1.5 + 5 + 1.5 = 21$

$\Sigma (R_1)^2 = 21^2 = 441$

$\Sigma R_2 = 7 + 10.5 + 8 + 6 + 3 = 34.5$

$\Sigma (R_2)^2 = 34.5^2 = 1190.25$

$\Sigma R_3 = 14 + 10.5 + 15 + 12.5 + 12.5 = 64.5$

$\Sigma (R_3)^2 = 64.5^2 = 4160.25$

Step three: Calculate $H_{obtained}$ using the formula.

$$H_{obt} = \left[\frac{12}{N(N+1)}\right]\left[\sum\frac{(R_i)^2}{n_i}\right] - 3(N+1)$$

Note that the capitalized "N" is the total N for all subjects participating the experiment, regardless of condition, while the small "n_i" refers to the sample size in each group/condition respectively. Confusing these two values is a common error, so be very careful as you work through examples on your own that you keep track of which sample size is required in the formula.

$$H_{obt} = \left[\frac{12}{15(15+1)}\right]\left[\sum\frac{(21)^2}{5} + \frac{(34.5)^2}{5} + \frac{(64.5)^2}{5}\right] - 3(15+1) =$$

$H_{obt} = (0.05)(88.2 + 238.05 + 832.05) - 48 =$

$H_{obt} = 57.915 - 48 = 9.915$

Step four: Evaluate $H_{obtained}$ using the Chi-square table

You need to know the number of groups in the experiment (k), and your α level in order to look up the H critical value. In this example you have three conditions, so the df = k-1 or 3-1=2, and $\alpha = 0.05$. Under these conditions, your H critical value is 5.991.

Decision Rule: If $H_{obtained} \geq H_{critical}$, then reject the null hypothesis.

Since $9.915 \geq 5.991$, we can reject the null hypothesis. The three conditions vary significantly in GPA scores, suggesting that time of day influences course performance.

Results if you use SPSS to conduct the Kruskal-Wallis Test
NPar Tests
Kruskal-Wallis Test

Ranks

	CLASS	N	Mean Rank
GPA	1.00	5	4.20
	2.00	5	6.90
	3.00	5	12.90
	Total	15	

Test Statistics

	GPA
Chi-Square	9.968
df	**2**
Asymp. Sig.	**.007**

a Kruskal Wallis Test
b Grouping Variable: CLASS

Note that SPSS reports a Chi-Square obtained value rather than an H obtained value. In addition, the value is adjusted because we have a small sample size. However, SPSS uses the same degrees of freedom (based on Chi-Square test) that we used and calculates an exact probability (0.007) that is statistically significant, thus the result is that same – we must reject the null hypothesis.

How would these results be reported in a scientific journal article?

If the journal required American Psychological Association (APA) format, the results would be reported in a format something like this:

> The three conditions vary significantly in GPA scores, suggesting that time of day influences course performance, $H\ (N=15) = 9.92, p < .01$.

This formal sentence includes the dependent variable (GPA), the independent variable (time of day in three categories), as well as a statement about statistical significance, the symbol of the test (H), the sample size (15), the statistical value (9.92), and the estimated probability of obtaining this result simply due to chance (<.01).

Assumptions for the Kruskal-Wallis (H) test

1. The dependent variable (data) are measured on an ordinal or better scale.
2. The scores come from the same underlying distribution.
3. You must have at least 5 scores per group to use the chi square table.
4. Random sampling

Spearman's rank order correlation (rho, or r_s)

The Spearman rank correlation test is appropriate when one or both of the variables of interest are on an ordinal scale. This test replaces the Pearson's correlation test when the data are not collected on an interval or ratio scale.

Calculation steps

1. Independently rank each variable X and Y (strongest response = 1).

2. Give tied ranks an average ranking score, but then don't reuse the ranks that you averaged. This is the same technique that we used for the Mann-Whitney U test and Wilcoxon when we had tied ranks.

3. Calculate the "difference in rank" scores [$D_i = R(X_i) - R(Y_i)$] and the squared differences in rank.

4. Calculate rho (r_s) using the formula:

$$r_s = 1 - \frac{6(\sum D_i^2)}{N^3 - N}$$

5. Evaluate r_s in the same way that we evaluated Pearson r.

(9) **Complete Example**

Determining the correlation between high intelligence (IQ scores on an interval scale) and a high level of aggressiveness of an individual (measured on an ordinal scale from 1-4 with 1 being the most aggressive) can be analyzed with the Spearman Rank Order Correlation, or rho. It is necessary to use the Spearman Rank Order Correlation, or rho, rather than the Pearson r Correlation because one of our variables is measured on an ordinal scale. Actually, when there are very few ties in the data, the equation we use for Spearman Rank simplifies to the Pearson r equation (except that we have ranked the scores to fit the assumptions).

Data

X (Aggressive Personality Scale)	Y (IQ Score)
4	92
3	128
4	134
2	98
1	87
1	101
3	100
2	108

Step One: Rank the Two Variables Separately and Handle Tied Ranks Appropriately

Xi (Aggression)	$R(X_i)$	Yi (IQ Score)	$R(Y_i)$
4	**7.5**	92	7
3	**5.5**	128	2
4	**7.5**	134	1
2	**3.5**	98	6
1	**1.5**	87	8
1	**1.5**	101	4
3	**5.5**	100	5
2	3.5	**108**	3

Note that the lower scores on the aggressiveness scale mean that the individual is more aggressive, so they get top ranks (one and two, etc.). Similarly, the higher IQ scores imply greater intelligence, so a high IQ score receives the top rank (rank = 1 for an IQ of 134). If you were to mix these, then you'd be trying to correlate low intelligence with high aggressiveness or high intelligence with low aggressiveness. In this case we are specifically trying to evaluate any potential correlation between high IQ and high aggressiveness, so high values for either variable should receive top ranks.

Step Two: Calculate "difference in rank" scores [$D_i = R(X_i) - R(Y_i)$] and square the difference scores as a intermediate step in preparation for using the r_s formula

X_1 (Aggressive)	$R(X_1)$	X_2 (IQ Score)	$R(X_2)$	D_i	D_i^2
4	**7.5**	92	**7**	**0.5**	0.25
3	**5.5**	128	**2**	**3.5**	12.25
4	**7.5**	134	**1**	**6.5**	42.25
2	**3.5**	98	**6**	**-2.5**	6.25
1	**1.5**	87	**8**	**-6.5**	42.25
1	**1.5**	101	**4**	**-2.5**	6.25
3	**5.5**	100	**5**	**0.5**	0.25
2	**3.5**	108	**3**	**0.5**	0.25
				$\Sigma = 0$	$\Sigma = 110$

Step Three: Calculate rho (r_s) using the formula below

$$r_s = 1 - \frac{6(\sum D_i^2)}{N^3 - N}$$

$$r_s = 1 - \frac{6(110)}{8^3 - 8} = 1 - \frac{660}{504} = -0.3095 \text{ or } -0.31 \text{ (rounded to two decimal places)}$$

Step Four: Interpreting the Correlation Coefficient

The Spearman rank order coefficient (r_s) coefficient for our IQ-aggressiveness example data is -0.31. A correlation coefficient this size suggests a fairly weak NEGATIVE relationship between our two variables. But what we really need to do is to determine the probability that we could get a statistic of -0.31 or more extreme by chance, the same question that we have been asking throughout much of Unit 2 and 3. We also need to keep in mind that the negative correlation suggests that top ranking IQ scores are associated with bottom ranking aggressiveness scores. This effect is in the OPPOSITE DIRECTION OF THE RESEARCHER'S ORIGINAL HYPOTHESIS. One could argue that these researchers should have used a two-tailed hypothesis and analysis. Evaluating all hypotheses as two-tailed comparisons protects the researcher from instances where the real effect is in the opposite direction that they predicted.

For the Pearson's r, we were determining the probability that, given a sample coefficient of -0.31, the correlation coefficient of the underlying population is really zero, reflecting no relationship between X and Y. We evaluated the Pearson r value using a table for critical values of r based on the degrees of freedom (N-2) and the desired level of alpha (see Table E). We will evaluate the Spearman rank order coefficient in the same way. In our case, the df are equal to 6 (8 - 2 = 6) and $\alpha = 0.05$. The one-tailed critical value is 0.6215 while the two-tailed critical value is 0.7067. Thus, whether or not you evaluate a one or two-tailed test does not change the outcome. The absolute values of the r_s obtained (0.31) calculated in this example is less than the two-tailed critical value (0.7067) AND the obtained value is not more extreme than the one-tailed critical value (0.6215) particularly when you consider that the critical value is positive and your obtained value is in the opposite tail of the distribution.

Another way to express the relationship between the independent and the dependent variables is to square your r-obtained value to make it R^2.

$$r_s = \sqrt{proportion - of - total - variance - of - Y - that - is - explained - by - X}$$

R^2 = The proportion of the variability in Y that is explained by X.

The use of the capitalized R is intentional as R^2 is the formal symbol for this statistic. If we apply this concept to our example, then R^2 is the proportion of the variance in intelligence that is explained by how aggressive a personality the individual exhibits. Obviously there are other possible (and potentially more likely!) factors that influence intelligence. If $r_s = -0.3095$, then $R^2 = (-0.3095)^2 = 0.09579$. This means that approximately 9.579% of the variability in intelligence is explained by aggressiveness in the individual's personality. If this were the case, then only 9.579% of the variability in your dependent variable would be explained by your independent variable, with 90.421% of the variability in your dependent variable (Y) left unexplained. This would suggest that the factor you are studying does not have a strong relationship with your dependent variable.

Results of the r_s correlation if you use SPSS

Correlations

			AGGRESS	IQ
Spearman's rho	AGGRESS	Correlation Coefficient	1.000	.342
		Sig. (2-tailed)	.	.408
		N	**8**	**8**
	IQ	Correlation Coefficient	.342	1.000
		Sig. (2-tailed)	.408	.
		N	**8**	**8**

Note that SPSS has calculated a slightly different value for Spearman's rho (r_s) at 0.342. This is most likely due to the way that SPSS handles tied ranks, and the bottom line is that the value is similar to our calculated value (although SPSS does not rank scores in a manner that matches your hypothesis so you need to be clear on the relationship between the two variables). Clearly, both our manual analysis and the computerized analysis agree that the results are not significant (SPSS assigns the significance or p-value at 0.408 which is greater than 0.05).

How would these results be reported in a scientific journal article?

If the journal required American Psychological Association (APA) format, the results would be reported in a format something like this:

> There was no significant relationship between high IQ and a high level of aggressiveness in an individual, $r_s(6) = -.31, p > .05; R^2 = .10$.

This formal sentence includes the two variables (IQ score and aggressiveness score) as well as a statement about statistical significance, the symbol of the test (r_s), the degrees of freedom (6), the statistical value (-.31), and the estimated probability of obtaining this result simply due to chance (>.05).

Final Flowchart on Choosing the Appropriate Test

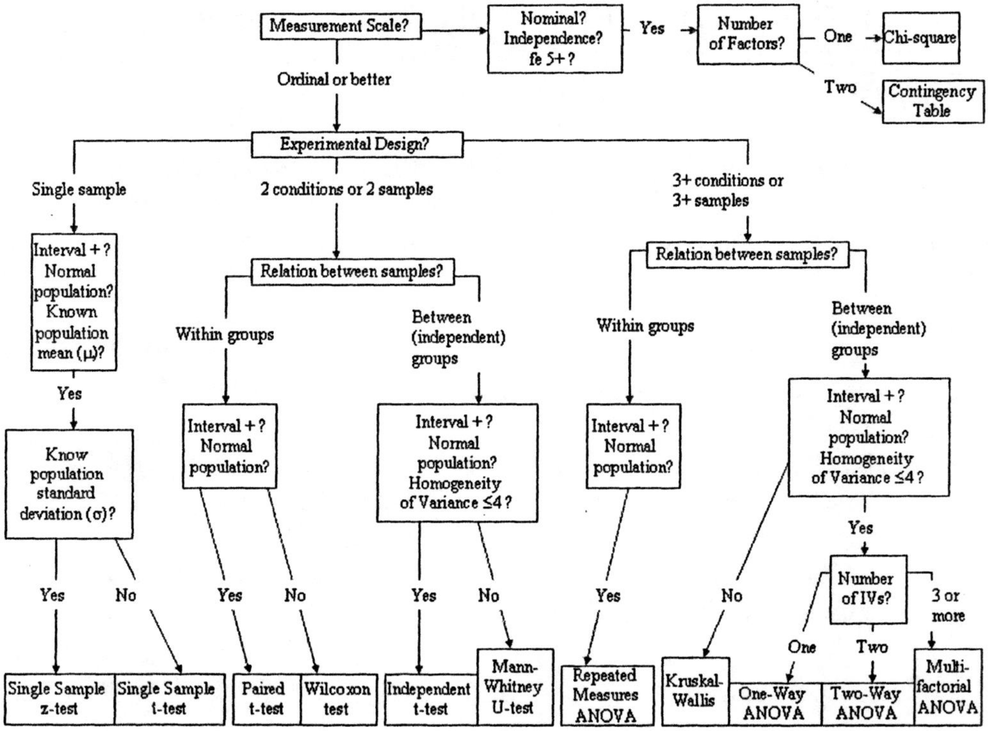

This flowchart includes the new tests that were introduced in this chapter and summarizes the inferential tests that you learned in this textbook. We have placed notes regarding the major assumptions of the test in the boxes as a reminder, but you must know the rules for meeting each assumption in order to use this flowchart effectively. We will review the assumptions for each test in the final chapter.

Summary

In this chapter you have covered statistical tests that are appropriate for use when you do not meet the assumptions for parametric tests. We have shown you the nonparametric equivalents of the paired/correlated t-test, the independent t-test, and the one-way ANOVA. In addition, we discussed the Chi-square and contingency table analyses for use with nominal data and the Spearman Rank Correlation Coefficient that is appropriate when one or more of the variables are ordinal. In the next chapter, we will review the tests you have learned in this book.

Chapter Twelve Homework

Provide a short answer for the following questions.

1. If there is a large difference between f_o and f_e for each cell, are you likely to reject or fail-to-reject the null hypothesis?

2. What are the assumptions for the χ^2 (Chi Square) test?

3. What are the assumptions for the Wilcoxon test?

4. What are the assumptions for the Mann-Whitney U test?

5. What are the assumptions for the Kruskal-Wallis Test?

6. When is it appropriate to use a Spearman's Rank test?

7. What do you do if f_e is less than 5?

8. What is a contingency table?

9. List three parametric tests and their non-parametric equivalent tests.

10. Chi-square is used to test differences between __________
    a. frequencies
    b. means
    c. variances

11. A survey was conducted to determine whether there was a relationship between age (25 and under or over 25) and preference for Coca Cola and Pepsi. The following data were recorded. Using an alpha level = .05, what do you conclude?

Soda Preference

	Pepsi	Coca Cola	Total
≤ 25 years	82	118	200
> 25 years	68	132	200
Total	150	250	400

12. A manufacturer of watches is examining preferences for digital verses analog watches. A sample of 200 people is selected and are classified for age and preference. Is there a relationship between gender and watch preference? Use an alpha = 0.05.

	Digital	Analog	Undecided	Total
Male	90	40	10	
Female	10	40	10	
Total				

13. A random sample of 50 severely ill patients from a medical center were asked whether they were satisfied with their doctors. Is there a significant difference between the observed data and the expected data? Use an alpha of 0.01.

	Satisfied
Very	15
Not at all	35

14. Ten years ago the Generic School District conducted district-wide reading assessments in grades 3, 6 and 9. The results indicated that 62% of 3rd graders, 56% of 6th graders and 42% of 9th graders performed satisfactorily on the assessment. The assessment was repeated a decade later after new reading programs were established

and 400 students in each grade level took the assessment. Has the performance changed in the last 10 years?

	3rd	6th	9th
# with satisfactory scores	298	305	217

CHOOSE THE MOST APPROPRIATE AND POWERFUL TEST:

15. A developmental psychologist is examining problem solving ability for grade school children. Random samples of five year old, six year old, and seven year old children are obtained with three children in each sample. Each child is given a standardized problem-solving task, and the psychologist records the numbers of errors.

16. Five pairs of teenagers are randomly selected and matched on a measure of general fitness. Each member of each pair is then randomly assigned to either the walking treatment or the dance treatment. After a specified time, students in each group are given a test of cardiovascular fitness. Heartbeats per minute are obtained for each subject and analyzed.

17. A researcher would like to measure the effects of air pollution on life expectancy. Two samples of newborn rats are randomly selected from a normally distributed population. The first sample of 10 rats is housed in cages where the atmosphere is equivalent to the air in a polluted city and lives for an average of 474 days (s=32). The second sample of 20 rats is placed in clean air cages and lives for an average of 556 days (s=4). Does pollution cause a difference in life expectancy?

18. A researcher examines the effect of relaxation training on test anxiety. Baseline levels of test anxiety are recorded for a sample of 31 subjects.

All the subjects then receive relaxation training for three weeks and their anxiety levels are measured again. Is there a treatment effect?

Chapter Thirteen
Review

In this chapter we review the concepts and statistical tests from the previous chapters. We have organized the review so that we cover the terms and concepts on a chapter-by-chapter basis. Our goal in this section is to integrate the terms from earlier chapters with the terms and concepts in later chapters. There are connections that between chapters that can only be made now that you have been exposed to all of the material in this book. In the second section of this chapter, we address how to choose the appropriate statistical analysis for the independent and dependent variables in a particular study. This is critical as most software programs do not do this for you, and your ability to evaluate the statistical analyses of others is dependent on your skills to determine which analysis matches a given experimental design. Lastly, we present a real-world example of a data set and demonstrate how to conduct the appropriate analyses and interpret the results using SPSS. We have deliberately chosen a data set that is conducive to more than one analysis, which is typically the case with data sets that are from large studies and are worthy of publication. In other words, more real-world behavioral data sets are extensive so that the researchers can attempt to address more than one hypothesis or question.

Terms-and-concepts review

Chapter Two

This chapter introduced many of the specialized terms that we use in statistics. Here you were first exposed to the concepts of a **population** and a **sample**, which laid the groundwork for recognizing single sample designs and determining whether a correction factor was required when calculating the standard deviation. We described why we must divide by n-1 when estimating the population standard deviation from a sample, but we divide by N when calculating the standard deviation of a population (where all scores of interest were known). By now, it should be clear that understanding the differences between a sample and a population is critical to becoming competent in statistics.

Another important early concept was that of a **variable** (any property that may take on different values at different times and may change with various conditions). Scientists use both **independent** and **dependent variables** in their research, and independent variables are those that we usually manipulate or control while dependent variables are those that we measure. Variables can either be **continuous** (have an infinite number of values between each unit) or **discrete** (have no values between each unit). Independent variables that are continuous (such as Age or IQ) are appropriate for correlation or regression analyses while independent variables that are discrete (Male or Female) are commonly analyzed with t-tests and ANOVA. Dependent variables can also be discrete (**nominal** or **ordinal scale**) or continuous (**interval** or **ratio scale**) and this impacts the decision about the appropriate analysis to conduct. For this reason, it's worth reviewing these scales again.

Four Measurement Scales

Nominal: data are on a categorical, and often qualitative, scale rather than one that is quantitative.

Ordinal: data are on a categorical scale, in which categories can be ranked in relative order.

Interval: data are collected in a manner that measures actual magnitude and which has equal intervals between possible scores, but does not have a meaningful absolute zero point.

Ratio: data are collected in a manner that measures magnitude, has equal intervals between possible scores, and contains an absolute zero point.

Chapter Three

In this chapter we introduced the concepts of **descriptive** (a summary of the data) versus **inferential statistics** (using sample statistics in an attempt to infer the characteristics of the population), and then covered a number of descriptive statistics in more detail. We introduced the three primary measures of **central tendency**, the **mean**, **median** and **mode**, and we discussed when each measure is most appropriate as a descriptive statistic.

The mean, however, is the measure that we used repeatedly in later chapters on inferential statistics (such as t-tests and ANOVA) because sample means tend to provide the most reliable estimates of population means. In other words, the mean is the measure of central tendency that is least subject to sampling variation. The mean is also useful since, from the mean, we can produce a measure of **standard deviation** and **variance**, which are both important for measuring the variability of scores in a sample or population. Recall that the standard deviation is the average deviation of a score from the mean, while the variance is the squared average deviation of a score from the mean.

Another important concept in this chapter was the issue of the shape of the distribution of scores. We defined a **symmetrical distribution** as one that was identically shaped on either side of the mean, or each side was the mirror image of the other. A distribution was skewed (not symmetrical) when there was a higher frequency of scores on one side of the mean compared to the other side of the mean. If there were more scores on the right hand side of the mean, then the distribution was negatively skewed. If there were more scores on the left hand side of the mean, then the distribution was positively skewed. When the **skewness** value of a distribution was zero, the distribution was not skewed (it was symmetrical). Another measure of the shape of a distribution is kurtosis. **Kurtosis** is another descriptive statistic, but one that measures whether the frequency of scores in a distribution is more or less heavily clustered around the mean compared to a normal distribution. A distribution with a kurtosis = 0 is normally distributed and not kurtotic. This discussion of symmetrical and bell-shaped distributions laid the foundation for your understanding of what a normal distribution should look like, and provided one way to determine whether the distribution was normally distributed. This concept became critical when we discussed assumptions requiring that our data be normally distributed.

Chapter Four

The critical information in this chapter was the continuing coverage on **normal distributions**, including applications of using a normal distribution to standardize scores (z) and to determine the probability associated with each score. The definition of a normal distribution and the concepts of probability are critical because many inferential tests require that your data be normally distributed so that the probability of obtaining a particular result by chance can be calculated. This was your first exposure to calculating statistics on your data and interpreting them based on the likelihood of obtaining that statistic simply due to chance, and not because of your experimental manipulations.

Chapter Five

In chapter five, we continued describing the process that one uses to evaluate whether a particular experimental result is due to the experimental manipulation or due to chance. We explained that this process is dependent upon **random sampling**; random sampling, however, is not a guarantee that we will always make the correct decision about the meaning of our experimental results. We can make mistakes, and those mistakes come in two different forms. We may unknowingly make a **Type I** (reject the null hypothesis when it is in fact true) or **Type II error** (fail-to-reject the null hypothesis when it is in fact false), but most of the time our experiment and our analysis should be done carefully enough that we should make correct decisions. Traditionally, we establish a maximum limit to the probability of making a Type I error by establishing **alpha**, generally at $p = 0.05$, before we begin our analyses.

If your independent variable truly affects your dependent variable, then you have a "**real effect**" of your independent variable. When this occurs, you should be able to reject your null hypothesis (which said there was no effect). If you cannot do so, it is because your experiment lacks power. **Power** is a measure (between 0 – 1.00) of the ability of a statistical test to detect an effect of the independent variable if the effect exists in reality. Power is influenced by many factors including sample size, the size of your real effect, your experimental design, the type of statistical test, whether your hypothesis is one-tailed or two-tailed, and at what level you set alpha. **Parametric** tests (such as t-tests and ANOVA) are more powerful than **non-parametric** tests (such as Mann-Whitney U and Wilcoxon), but parametric tests require that you make some assumptions about the underlying population parameters. If you do not meet those assumptions, the results of parametric analyses conducted on those data are suspect. Many scientific journals and funding agencies (such as the National Institutes of Health and the National

Science Foundation) require that you estimate the power of your experiment so that reviewers can better interpret your findings. Knowing that your power is 0.92 versus 0.22 is an important factor in interpreting your possible failure to reject the **null hypothesis**.

Chapter Six

In chapter six, we introduced our first hypothesis tests: those that test hypotheses having to do with **single sample** designs when compared to underlying populations. We began with a simple application of the concept of standardized z-scores, which became the single subject z test. We then introduced designs which compare a sample mean to a population mean, rather than a single score to a population mean, and developed the single sample z test. We went on to make this concept more useful by relaxing the (admittedly uncommonly satisfied) assumption of a known population standard deviation by allowing the estimation of the population standard deviation from the sample standard deviation: the single sample t test. This required a different **sampling distribution** for calculating the resulting probabilities of the null hypothesis, and we introduced the t distribution, which is influenced by sample size, or more specifically, degrees of freedom. The concept of a sampling distribution was elaborated in this chapter. We defined a sampling distribution as a hypothetical representation of all possible values of a statistic with the probability of each possible value if there is no effect of an independent variable. This further reinforced earlier concepts that suggested that we are always comparing our actual statistic to the probability of obtaining that statistic randomly. Evaluating this likelihood requires an appropriate sampling distribution for that statistic. Finally, we applied the single sample t test to a very useful descriptive statistic called the **confidence interval**. The confidence interval for the population mean is a range of values with a calculated probability of containing the mean.

Chapter Seven

Here we were finally able to cover statistical tests for designs in which you have two samples and you lack information about the underlying population. Because we rarely know population parameters such as the mean and standard deviation (μ and σ), the two sample tests tend to be used more often than the single sample tests, but they are also less powerful. They are less powerful than single sample tests because we are forced to estimate characteristics of the population whereas single sample tests rely on known measures of the population. Nevertheless, independent and correlated (or paired) t-tests are some of the most commonly used parametric statistics in science.

Chapter Eight

In this chapter we covered statistical tests for designs in which you have three or more samples in an independent, or between-groups, design, and you lack information about the underlying population. You were shown the steps involved in conducting the overall analysis (F test), as well as the steps involved in determining which means are significantly different (providing that your first overall F test is significant). An F test (or ANOVA) is more appropriate than performing multiple independent t-tests for the 3+ groups/conditions as it controls the probability of **Type I error** by looking for one overall difference between all of the means prior to moving on to group-by-group comparisons. An F-obtained value is determined by dividing the **within groups variance** (reflecting natural variation in the population, thus an estimate of σ^2) by the **between groups variance** (reflects natural variation in the population AND the effect of the independent variable, if any).

Chapter Nine

This chapter focused on more complex types of ANOVA where you have a within-groups design or a between-groups (independent) design with multiple independent variables. Designs with multiple independent variables result in F-obtained values that are produced for each independent variable (or main effect) as well as F-obtained values for the potential interaction of the independent variables. Recall that a **main effect** is the influence of your independent variable on your dependent variable, while an **interaction** is the (possibly different) effect of one of your independent variables (or factors) at different levels of your other independent variable (or factor).

Chapter Ten

Linear regression and correlation are closely related, but there are some differences between them to help us distinguish which is most appropriate for a given situation. Linear regression is used for prediction purposes and calculates an equation for the "best-fit" line which answers the question, "what is the relationship between X and Y". Correlation analysis focuses on the strength of the relationship between X and Y, or how tightly the data points are clustered around the "best-fit" line. Both linear regression and correlation are appropriate when both variables being measured are continuous rather than categorical. Relationships between the variables may be positive (direct), negative (inverse), perfect or imperfect. Positive relationships are those where Y increases as X increases, while negative relationships show that Y decreases as X increases. Perfect relationships are those where all data points fall on the best-fit regression line, whereas imperfect relationships occur when not all data points fall on this line. Hence, imperfect relationships are more common than perfect relationships (and not just in statistics!).

Chapter Eleven

The modern **general linear modeling** (GLM) approach is very powerful, allowing the simultaneous assessment of continuous and categorical independent variables. Thus, researchers can combine multiple independent variables (or predictors) that traditionally would have been analyzed separately as regression (continuous variables) and ANOVA (categorical variables) into one statistically powerful and complete overall analysis. This is advantageous for multiple reasons, including developing complete models to explain behavior as well as limiting the probability of a Type I error. The concept of general linear modeling has been widely embraced by the research community but is still rarely included in undergraduate statistics textbooks. More importantly for this course, these concepts help us to interpret the results of statistical software packages that use the GLM terminology, which has become universal.

Chapter Twelve

In this chapter, you discovered statistical tests that are appropriate for use when you do not meet the assumptions for parametric tests. We have shown you the nonparametric equivalents of the paired/correlated t-test, the independent t-test, and the one-way ANOVA. In addition, we discussed the Chi-square and contingency table analyses for use with nominal data and the Spearman Rank Correlation Coefficient that is appropriate when one or more of the variables are ordinal. In general, these tests are not as powerful as the parametric tests, but they are useful when the assumptions of parametric tests are not met. The assumptions can be broken due to non-normality of the data, lack of homogeneous variances, or because the researcher designed an experiment where ordinal scale data was collected. This is commonly done in many areas of behavioral research, primarily in the form of questionnaires where subjects are asked to respond to specific questions on an ordinal scale. For example, subjects might be asked to rank their preference of a soft drink from 1 to 5, with 5 suggesting their highest preference. There is some indication that parametric tests can be performed on ordinal data, but this issue is hotly debated.

Choose-the-appropriate-test review

So far you have primarily used the flowchart to determine the appropriate test for an analysis. Some of our undergraduate statistics students have also found an outline or table format a useful way to organize the information in the flowchart. In Table 13.1 we have organized the statistical tests that were covered in this book based on the experimental design or measurement scale (our outline format). If you include the post hoc test (Newman-Keuls), you have learned ten inferential tests in addition to correlation, regression and descriptive statistics.

Table 13.1
Outline Format for Choosing the Appropriate Inferential Test

Single Sample Designs
- *z-test for single samples*
- *t-test for single samples*

Two Groups/Paired or Correlated
- *Paired t-test*
- *Wilcoxon signed ranks test*

Two Groups/Independent
- *Independent t-test*
- *Mann-Whitney U-test*

Multigroup/Independent & Single Factor
- *One-way Analysis of Variance (ANOVA)*
    - A priori or planned comparisons
        - Independent t-test
    - A posteriori or post hoc comparisons
        - Newman-Keuls
- *Kruskal-Wallis H-test*

Nominal data
- *Chi-Square*

Table 13.2 presents the choose-the-appropriate test criteria in a true "table format" with the typical assumptions of the parametric tests numbered from 1-5. Note that some assumptions for the chi-square and the assumption of random sampling are not specifically addressed in this brief decision table format.

Table 13.2
Table Format for Choosing the Appropriate Inferential Test

# of groups OR conditions	**Type of design**	E. <u>Assumptions</u> F. <u>(see numbered text)</u>	**Type of test to use**
1 sample	Single sample	1, 2, 4 & 5 are all met	Single sample z-test
1 sample	Single sample	1, 2, & 4 are all met	Single sample t-test
2	Independent (**Between**) groups	1, 2 & 3 are met	Independent t-test
2	Independent (**Between**) groups	At least 1 2, or 3 broken	Mann-Whitney U
2	Dependent (**Within**) groups	1 and 2 are both met	Paired t-test
2	Dependent (**Within**) groups	At least 1 or 2 broken	Wilcoxon
1 factor		Using nominal data	Chi-Square
2 or more factors		Using nominal data	Contingency Table
3 or more	Independent (**Between**) groups	1,2 & 3 are met	ANOVA
3 or more	Independent (**Between**) groups	At least 1, 2 or 3 broken	Kruskal-Wallis

Have these assumptions been met?

1. The data must be interval or ratio.
2. The data are normally distributed, meaning a) the population raw scores are known to be normally distributed, or b) the sample size is $\geq$ 30, or c) the skewness and kurtosis values are approximately between -1.0 and +1.0.
3. The variances are equal between the groups, called homogeneity of variance (HOV). The variances can be up to 4 times different from each other, but no more than that, and still be considered "equal." To find HOV, you divide the larger variance by the smaller variance.
4. Known population mean
5. Known population standard deviation

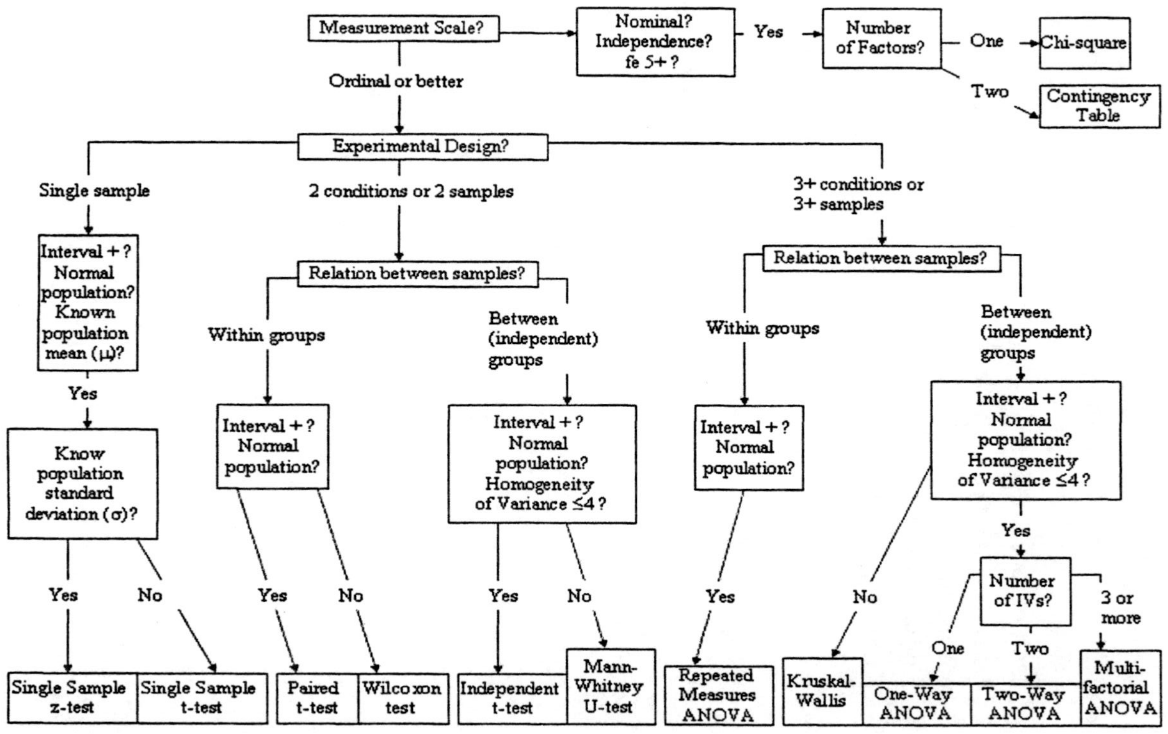

Figure 13.1 Choosing the Appropriate Inferential Test Flowchart

Figure 13.2
Flowchart for Choosing the Appropriate Inference Test When IV is Continuous

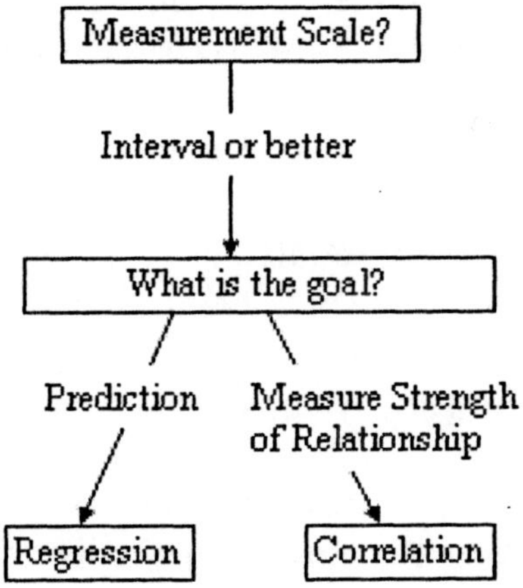

Tying-it-all-together review

Real-world research often involves more than one hypothesis, independent variable and dependent variable. Funding agencies, scientists, students and the general public typically want to maximize the amount of information that can be obtained from a single study. This makes sense from both an academic as well as a financial perspective. For pedagogical reasons, we have chosen to minimize the complexity of many of the experimental designs and datasets that were presented in this book. Now that you are more familiar with statistics, you are ready to begin familiarizing yourself with more complex designs and analyses. Here we present a research design and data that are appropriate for more than one analysis, along with the SPSS output from those analyses.

Real-World Example

Research Questions and Design

An organizational psychologist examined the factors related to burnout or stress in engineers. Data were gathered from volunteers (n = 50) chosen by randomized selection from a professional technical union. The first hypothesis was that stress would be positively related to age. Exact age was determined by birthdate. The second hypothesis was that gender and marital status (Married or Unmarried) would be related to the emotional exhaustion level of burnout/stress. Stress was measured as the level of cortisol (a hormone associated with stress) in the bloodsteam (pic/ml) and as the number of days of work missed due to illness.

Determining the Appropriate Analyses

Based on the story, hypothesis 1 is completely separate from hypothesis 2. Hypothesis 1 could be analyzed as a correlation because both variables (age and stress) have been measured on a continuous scale. In fact, we can perform two correlational analyses as there are two dependent variables for stress (cortisol and days sick), but only one independent variable (age). Hypothesis 2 is more complex, with two independent variables that are categorical (Gender and Marital Status) in addition to the two dependent variables (cortisol and number of days sick). In this case, it is appropriate to use the General Linear Modelling function in SPSS or a similar statistical package.

SPSS Output and Interpretation for Hypothesis One

Exploring the Relationship between Cortisol and Age in Engineers

Correlations

Correlations

		AGE	CORTISOL
AGE	Pearson Correlation	1.000	.391**
	Sig. (2-tailed)	.	.005
	N	50	50
CORTISOL	Pearson Correlation	.391**	1.000
	Sig. (2-tailed)	.005	.
	N	50	50

**. Correlation is significant at the 0.01 level (2-tailed).

It appears that age (in years) and cortisol (pic/ml) in the blood stream are positively correlated. Note that the Pearson Product Moment Correlation value is 0.391 (a positive number) and that the two-tailed significance of this value is 0.005. This suggests a moderate correlation between age and cortisol.

Graph

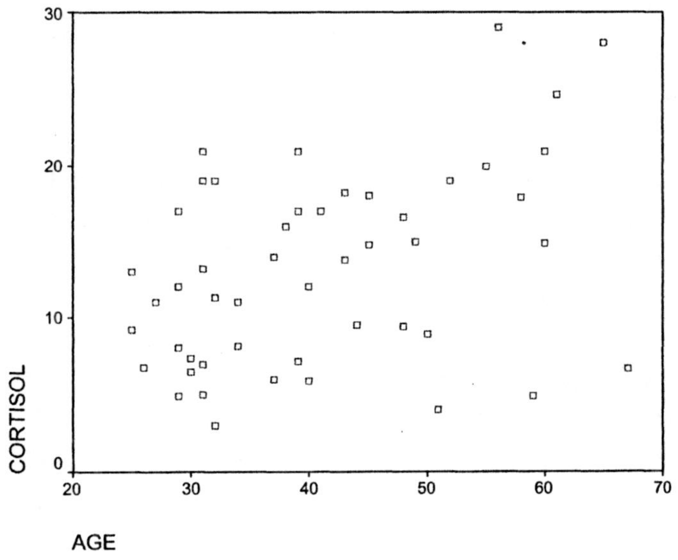

The simple scatterplot suggests a pattern that cortisol levels tend to increase with age. Of course, we must be careful to remember that correlation does not imply causation and that we cannot generalize beyond our population of interest (engineers in unions).

Exploring the Relationship between Sick Days and Age in Engineers

Correlations

		AGE	DAYSSICK
AGE	Pearson Correlation	1.000	.144
	Sig. (2-tailed)	.	.317
	N	50	50
DAYSSICK	Pearson Correlation	.144	1.000
	Sig. (2-tailed)	.317	.
	N	50	50

It appears that age (in years) and sick days (annually) are not related. Note that the Pearson Product Moment Correlation value is 0.144, a positive number but one in which the two-tailed significance is 0.317. This suggests no correlation between age and sick days.

SPSS Output and Interpretation for Hypothesis Two

General Linear Model

Between-Subjects Factors

		Value Label	N
GENDER	1.00	Male	34
	2.00	Female	16
MARITAL	1.00	Married	29
	2.00	Unmarried	21

SPSS automatically prints out your sample sizes and reminds you how you coded your data (1=Male and 2=Female, etc.).

Multivariate Tests[b]

Effect		Value	F	Hypothesis df	Error df	Sig.
Intercept	Pillai's Trace	.838	116.585[a]	2.000	45.000	.000
	Wilks' Lambda	.162	116.585[a]	2.000	45.000	.000
	Hotelling's Trace	5.182	116.585[a]	2.000	45.000	.000
	Roy's Largest Root	5.182	116.585[a]	2.000	45.000	.000
GENDER	Pillai's Trace	.021	.476[a]	2.000	45.000	.624
	Wilks' Lambda	.979	.476[a]	2.000	45.000	.624
	Hotelling's Trace	.021	.476[a]	2.000	45.000	.624
	Roy's Largest Root	.021	.476[a]	2.000	45.000	.624
MARITAL	Pillai's Trace	.044	1.024[a]	2.000	45.000	.368
	Wilks' Lambda	.956	1.024[a]	2.000	45.000	.368
	Hotelling's Trace	.045	1.024[a]	2.000	45.000	.368
	Roy's Largest Root	.045	1.024[a]	2.000	45.000	.368
GENDER * MARITAL	Pillai's Trace	.229	6.689[a]	2.000	45.000	.003
	Wilks' Lambda	.771	6.689[a]	2.000	45.000	.003
	Hotelling's Trace	.297	6.689[a]	2.000	45.000	.003
	Roy's Largest Root	.297	6.689[a]	2.000	45.000	.003

a. Exact statistic

b. Design: Intercept+GENDER+MARITAL+GENDER * MARITAL

The Notes that SPSS has placed under the Multivariate Tests box includes "b. Design: Intercept + Gender + Marital + Gender*Marital." This is the same format for the general linear model that we introduced in the GLM chapter. Both Gender and Marital are main effects, while Gender*Marital is the interaction. This table provides a multivariate test (overall test) of both of the dependent variables simultaneously and suggests that there are no significant main effects of gender or marital status, but there is a significant interaction of gender and marital status on overall stress (measured by cortisol and days sick). SPSS has conducted the analysis using <u>four</u> different multivariate techniques (Pillai's Trace, Wilks' Lambda, Hotelling's Trace & Roy's Largest Root). The subtle differences between these multivariate tests would be a topic for a more advanced course.

Tests of Between-Subjects Effects

Source	Dependent Variable	Type III Sum of Squares	df	Mean Square	F	Sig.
Corrected Model	CORTISOL	502.260[a]	3	167.420	5.354	.003
	DAYSSICK	125.081[b]	3	41.694	2.028	.123
Intercept	CORTISOL	7029.801	1	7029.801	224.792	.000
	DAYSSICK	380.436	1	380.436	18.504	.000
GENDER	CORTISOL	30.285	1	30.285	.968	.330
	DAYSSICK	.261	1	.261	.013	.911
MARITAL	CORTISOL	65.250	1	65.250	2.087	.155
	DAYSSICK	6.707E-03	1	6.707E-03	.000	.986
GENDER * MARITAL	CORTISOL	256.424	1	256.424	8.200	.006
	DAYSSICK	101.236	1	101.236	4.924	.031
Error	CORTISOL	1438.535	46	31.272		
	DAYSSICK	945.739	46	20.560		
Total	CORTISOL	10559.589	50			
	DAYSSICK	1539.000	50			
Corrected Total	CORTISOL	1940.794	49			
	DAYSSICK	1070.820	49			

a. R Squared = .259 (Adjusted R Squared = .210)

b. R Squared = .117 (Adjusted R Squared = .059)

SPSS also includes a Between-Subjects Effects table that shows the effects of each dependent variable separately from the other. Again, you should note that the main effects are not significant, but there is a significant interaction of gender and marital status on cortisol (significance or p-value = .006) and a significant interaction of gender and marital status on the number of days you call in sick (significance or p-value = .031). Interpreting the interaction would be facilitated by graphing the results and/or evaluating the descriptive statistics. Look over the means and standard deviations below to see if that helps you understand why there is no significant difference due to gender or marital status, but there is an interaction.

Descriptive Statistics

	GENDER	MARITAL	Mean	Std. Deviation	N
CORTISOL	Male	Married	10.1353	6.4893	19
		Unmarried	17.6187	4.6383	15
		Total	13.4368	6.8060	34
	Female	Married	13.4000	5.9666	10
		Unmarried	10.9350	3.4335	6
		Total	12.4756	5.1777	16
	Total	Married	11.2610	6.4038	29
		Unmarried	15.7090	5.2516	21
		Total	13.1292	6.2935	50
DAYSSICK	Male	Married	4.6842	5.7644	19
		Unmarried	1.5333	2.4456	15
		Total	3.2941	4.8149	34
	Female	Married	1.4000	2.2211	10
		Unmarried	4.5000	6.6257	6
		Total	2.5625	4.4717	16
	Total	Married	3.5517	5.0468	29
		Unmarried	2.3810	4.1289	21
		Total	3.0600	4.6748	50

While the pattern is fairly clear from the descriptive statistics, note how the graphs make the picture even clearer. Recall that crossing lines suggest an interaction!

CORTISOL

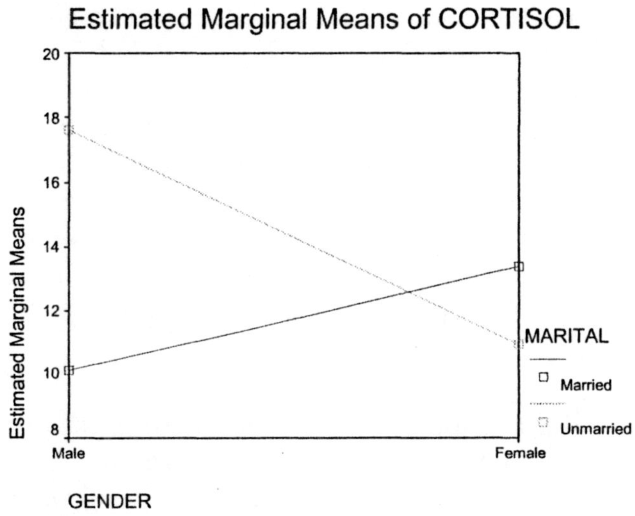

DAYSSICK

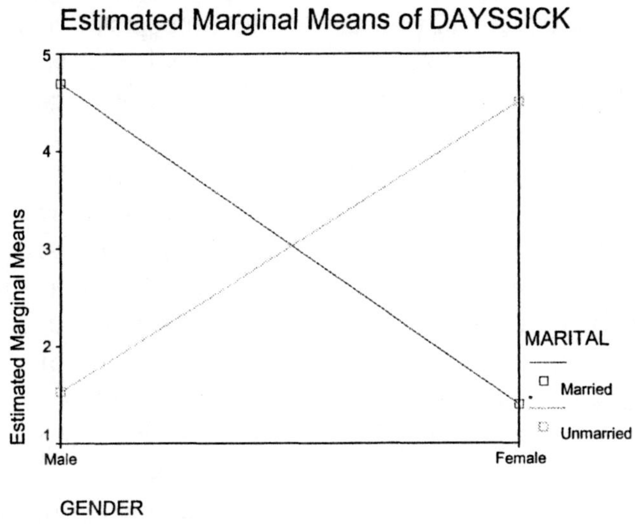

Conclusions from the Sample Data

These results suggest that there is a positive relationship between age and cortisol (r_{50}=.39, p<.01) but not between age and days called in sick

(r_{50}=.14, p=.32). Additionally, there is a significant interaction between gender and marital status on cortisol ($F_{1,46}$ = 8.20, p <.01), but no significant main effects of gender ($F_{1,46}$ = .97, p =.33)or marital status ($F_{1,46}$ = 2.09, p =.15). There was also a significant interaction between gender and marital status on the number of days called in sick ($F_{1,46}$ = 4.92, p =.03), but not for the main effects of gender ($F_{1,46}$ = .01, p =.91) and marital status($F_{1,46}$ = 0.00, p =.99). Specifically, male engineers show higher levels of cortisol when they are unmarried rather than married, but female engineers exhibit higher levels of cortisol when they are married. Male engineers are more likely to call in sick when they are married, but female engineers are more likely to call in sick when they are unmarried. Cortisol and the number of days called in sick do not appear to be measuring the same underlying stress processes.

Summary

In this chapter we reviewed the major terms and concepts from each of the following chapters, we presented the flowcharts for choosing the appropriate test including some new formats for that information, and we gave you an example of a complex research design and analysis that more closely resembles statistical analysis in the real world. In our own undergraduate statistical courses, we have found that students learn these concepts at very different rates. For some students the light bulb goes on immediately, but for others it is near the end of the course when the concepts begin to integrate and make sense. For still others, it is later, in advanced courses, when they truly understand the basic concepts and can internalize them fully. Regardless of where you fall in that continuum, you may find that you are able to return to the concepts presented in this book later with new understanding and insights. A colleague of ours once said that learning statistics is like walking up a spiral staircase. As you circle around and gain elevation, you find that you are looking at the same view, but with a new perspective. Statistics is one of those subjects where the more you learn, the more you understand the concepts that came before. As such, you may find this book is useful to you after you have completed your course in statistics. In fact, the books from my statistics courses (undergraduate and graduate) are some of the few textbooks I have kept over the years.

Chapter Thirteen Homework

Choose the most appropriate and powerful statistical test. Be sure to indicate any assumptions you may be making.

1. A researcher would like to measure the effects of a gene therapy on the life expectancy of roundworms. Two samples of roundworms are randomly selected from a normally distributed population. The first sample of ten worms is given no gene therapy and lives for an average of 24 days (s=8). The second sample of 20 worms is given a gene therapy and lives for an average of 33 days (s=3). Does the gene therapy extend the life of the worms?

2. A physician is testing the effectiveness of a new arthritis drug by measuring the grip strength of 10 patients before and after they receive the drug. Grip strength is measured on a 10-point scale, and the score for each patient is calculated as the difference in strength between the two conditions.

3. A psychologist interested in the development of manual dexterity prepared a block-manipulation task for three-year-old children. A sample of 83 children was obtained, with 40 girls and 43 boys. The psychologist recorded the amount of time (sec) required by each child to arrange the blocks in a specified pattern. The girls took an average of 48 seconds and the boys took an average of 39 seconds. The variance for the girls was 18 and the variance for the boys was 6.

4. Four wines are rated according to taste preference. Two hundred subjects are randomly selected to rate the wines. Each subject rates only one wine (50 subjects are randomly assigned to taste each wine). The rating scale is from 1 to 20, with 20 representing the highest possible score. Is there a significantly higher preference for any of the wines?

5. A dog trainer is interested in whether his puppy classes have an effect on the time it takes for his dogs to complete an obstacle course later in life. He randomly selects a sample of 60 puppies to join his class. When they are one year old, he records how long it takes each of them to complete a standardized obstacle course for the first time. He finds that it takes an average of 9.2 minutes (std dev = 2.3 min). He knows from extensive historical records that the average one-year-old dog takes 10.9 minutes (std dev = 3.2 min) to complete the standardized obstacle course for the first time. Do his puppy classes have an effect?

6. A researcher is interested in whether there is an effect of alcohol on bowling ability. She records the scores of 65 subjects both before and after they consume a specified amount of alcohol. The average score before the alcohol is 115 (s = 15) and the average score after the alcohol is 113 (s = 29).

7. A researcher suspects that color blindness is inherited by a sex-linked gene. This possibility is examined by looking for a relationship between gender and color vision. A sample of 1000 people is tested for color blindness, and then they are classified according to their sex and color vision status (normal, red-green blind, other color blindness). Is color blindness related to gender?

	Normal	Red-green blind	Other color blind
Male	320	70	10
Female	580	10	10

8. An investigator conducts an experiment involving the effects of three levels of a drug on memory. Subjects are randomly assigned to one of three conditions. A different drug level is administered in each condition. Memory is measured 10 minutes after each subject receives the drug, and the variances are homogeneous.

Group means:
1. (5 mg drug) = 1, n=45
2. (20 mg drug) = 4, n=45
3. (50 mg drug) = 10, n=48

9. Researchers compared individuals who had and had not taken college-level mathematics on their knowledge of high school algebra. The mean of students on the exam who did not take math in college was 71.5, while the mean on the exam for students who did take math in college was 79.6. The variances were homogeneous. What can we conclude from these results?

10. A sport psychologist wanted to find out if pro basketball player's free throw percentage in games would improve if they shot 31 additional free throws before each game. Players were matched individually with other players on game percentage. Sixty-six players participated in the experiment. Half of the players received the treatment effect (shooting 31 additional free throws before each game) and the other half served as controls. Researchers then compared game free throw percentage between the conditions.

11. Researchers wanted to see if there is a relationship between faculty and students on how much they like the tag line **LEARNING AT THE LEADING EDGE**. Researchers found that 50 students liked the slogan, while 15 did not and 35 had no preference. 45 faculty members liked the slogan, while 15 did not and 40 had no preference. The variances were homogeneous. How would you analyze the data?

12. A survey was conducted to find out how well Sociology graduate students liked learning Sociology at UW compared to where they received their undergraduate degrees. Twenty eight Sociology graduate students were surveyed. 12 preferred their undergraduate school, and 16 preferred UW. The data were mutually exclusive.

13. Researchers analyzed how long bartenders (n=200), sales people (n=350), and clerical workers (n=150) could hold their breath under water. Individuals were randomly selected from a normal population. The variances were .37, .26, and .12.

14. Researchers hypothesized that after an observer rat interacted with a demonstrator rat that had a lemon extract on its fur, the observer rat would prefer lemon over vanilla extract. To conduct their study they gave 39 observer rats an initial preference test between vanilla and lemon where they measured the time spent on each scented side of a testing apparatus. Then they allowed all of the observer rats to interact with a demonstrator rat smelling of lemon. Then they conducted a second preference test where they recorded the amount of time rats spent on the lemon side after the social interaction compared to the time they spent on the lemon side in their initial preference test. How can we analyze these data?

15. Researchers wanted to know if exam scores would change if a chemistry class was held in the afternoon compared to the morning. Researchers paired students in two sections, one in the morning and one in the afternoon. The section instructor, GPA, and other extraneous factors were controlled. The researcher recorded the scores in each section. How can researchers analyze the data?

Answers to Homework Exercises

Chapter Two Homework Answers

1. A population is the entire possible data set. A sample is a subset of that data.
2. A variable is any property that may have different values at different times, under different conditions. A constant is a fixed value.
3. An independent variable is the variable being manipulated or changed; it is the cause of an effect. A dependent variable is the variable being measured; it is the effect (i.e., it is the variable that you look at to determine whether your independent variable had an effect).
4. A statistic is a summary calculation on *sample* data. A parameter is a summary calculation on *population* data.
5. A continuous variable can have an infinite number of values between each unit. A discrete variable has no values between each unit.
6. Descriptive statistics are used to summarize your sample data with a single number, such as a mean or range, while inferential statistics are numbers that we calculate based on sample data, but that we are attempting to use to generalize to a larger population of data.
7. Random sampling (that is, selecting subjects with no bias) allows us to use the laws of probability on the data we collect. It also permits us to make decisions about how representative that data is of the population.
8. There are four measurement scales that are used to collect data, and which one you use determines the mathematical manipulations that you can make on the data later.

9.

What is the independent variable?	hours of sleep per night (4 or 8)
What is the population in this study?	38,000 U. Washington undergrads
What is the dependent variable?	lecture attendance
What is the sample?	350 undergrads
What statistic is used?	mean lecture attendance

10.

What is the independent variable?	Gender of the teacher
What is the population in this study?	California high schoolers
What is the dependent variable?	Test scores
What is the sample?	150 high schoolers from Santa Monica County
What statistic is used?	mean test scores

11. Nominal data are categorical. There is no numerical relationship on which to base any order or ranking.
12. Ordinal data are rank-ordered data, but which does not have equal intervals between data points or any measure of magnitude.
13. Interval data have equal intervals and magnitude between data points, but do not have an absolute zero.
14. Ratio data are like interval data, only with an absolute zero.
15. A. Nominal
16. C. Ratio, Interval
17. B. Ordinal: A 1-7 ranking about how helpful these questions are
18. a. 7, b. 303, c. 16343, d. 91809, e. 261, f. 315

Chapter Three Homework Answers

1. A statistic that measures the middle, or central tendency, of a set of scores and is calculated by taking the sum of the scores and dividing by the number of scores.
2. The value at which 50% of the scores fall below it.
3. The most frequent score in a set of scores.
4. A distribution that only has one mode, or most frequent score.
5. A distribution is symmetrical if it is identical on each side of the mean (mirror images).
6. While both are measures of the average dispersion of scores around the mean, the variance is in squared units while the standard deviation is not.
7. A positive skew occurs when one tail of the distribution is longer in the direction of higher scores, while a negative skew occurs when one tail of the distribution is longer towards the lower scores.
8. When you estimate the population standard deviation from a sample standard deviation, it tends to underestimate the true variability of the population. If you subtract one from your sample size, it provides an adjustment that ensures that your estimate of the population standard deviation is not too small. Thus, the sample standard deviation is considered a biased estimate of the population standard deviation, but the sample mean is an unbiased estimate of the population mean. If the sample mean were a biased estimate, it would require a correction, but it an unbiased estimate of the population mean and no correction is required.
9. If you subtract every score from the mean of the collective scores and then sum those deviations, it will add up to zero. This is logical since the mean should arithmetically be the middle score.
10. We divide by the number of scores so that you end up with the *average* deviation around the mean.
11. B. Negatively skewed.
12. A. Positively skewed.

13a. $\mu = 20.37$, $SS = 3455 - (163^2/8) = 133.87$, $\sigma = 4.09$

13b. $\overline{X} = 20.375$, s = 4.37
13c. $s^2 = 19.12$
14. $\overline{X} = 26.50$, s = 1.41
15a. 269/12 = 22.42
15b. average of two centermost points (17+19)/2 = 18
15c. 19
15d. 12
15e. 22.56
16a. 643/9 = 71.44
16b. s = 5.66
16c. $s^2 = 32.03$

Chapter Four Homework Answers

Give short answers to these conceptual questions.

1. A number between 0 and 1 that indicates the likelihood of an event.
2. 5/6 that the event will not happen or 1/6 that the event will happen.
3. 2/3 that the event will happen and 1/3 that the event will not happen.
4. Infinity – you would never reach that point because the curve is asymptotic to the horizontal axis.
5. Probabilities vary between 0 and 1.
6. A priori probability refers to a probability value deduced from reason alone, while a posteriori refers to an actual probability value that is calculated AFTER collecting the data.
7. At z-scores of –1 and +1 (corresponding to –1 and +1 standard deviation)
8. Z-scores are data points transformed into standard scores with common units. Those common units are expressed as standard deviations from the mean.
9. -1.49, 0.55, -0.08, -0.47, 1.49

10a. X =45.90

10b. z = 1.58, so the first subject did better because the z score is larger.

11a. 8.44

11b. -0.94

11c. 51.31 and 56.69

12a. 28.43%

12b. 24.99%

13a. 1.87

13b. 79.07

14. The most extreme 5% of the scores fall below 98.64 and above 161.36.

15. The raw score that lies at the 90th percentile is 615.20

Chapter Five Homework Answers

1. A real effect is the actual effect of the independent variable, which is not known (i.e., it might be masked by confounds).
2. Barring other factors (e.g., increasing N), power changes directly with alpha. For example, as alpha decreases, power decreases (because Power = 1 – β, and as alpha decreases, β increases).
3. Logistically, we usually have to sample without replacement, but our statistical inferences require sampling with replacement.
4. So that the data is representative of the underlying population, and therefore we can use the laws of probability to make inferences.
5. A sample is random if each possible sample of a given size has an equal chance of being selected and all members of a population have an equal chance of being selected into the sample.
6. Increasing N increases power.
7. Some factors include sample size (N), the type of statistical test used, the magnitude of the real effect, the experimental design, and the alpha level.
8. Decreasing alpha increases the probability of a type II error (β) – accepting the null hypothesis when it's actually false – because they are inversely related.
9. You could argue that it would be reasonable to set the alpha level higher in this situation given the real-world consequences. Minimizing type II errors may be more important in this case than minimizing type I errors, because you really don't want to retain the null (the drug has no effect) if it's false.
10. Parametric statistical tests require assumptions about the parameters of the population while non-parametric tests do not require assumptions about the population.
11. a. power increases

12 c. power decreases

13 c. .95

14 b. minimizes beta

15 a. decrease, increase

16 d. all of the above

17. marital therapy affects happiness

18. marital therapy does not affect happiness
19. reject H_0 : marital happiness affects happiness
20. married couples in New York City
21. Type I error because it is possible that marital therapy does not affect happiness.
22. Negative verbal feedback affects test performance.
23. Negative verbal feedback has no affect on test performance.
24. Negative verbal feedback had a significant effect on test performance, $p<.05$
25. Afternoon exam scores will be higher than morning exam scores.
26. There will be no difference between afternoon and morning exam scores, or exam scores will be higher in the morning.
27. Since the effect is in the direction opposite the original prediction, the null hypothesis is retained. Afternoon exams do not increase scores above morning exams.

Chapter Six Homework Answers

1. All the values that the statistic can take and the probability of getting each value under the assumption that chance alone is acting.
2. An actual or theoretical population of scores that would result if the experiment were done on the entire population and the independent variable had no effect.
3. It is used to generate the sampling distribution that allows us to test the validity of the null hypothesis.
4. All possible values of the mean of samples of size N, with the probability of getting each value, assuming that sampling is random from the null-hypothesis population.
5. Regardless of the shape of the population of raw scores, the sampling distribution of the mean approaches a normal distribution as sample size N increases.
6. Each sample mean can be considered an estimate of the mean of the raw score population. This estimate is not precise, and thus there is some sampling error associated with this estimate.
7. Single Sample t test
8. Three of the following: Interval or ratio data, Random sampling (assumed for all hypothesis testing), Normal distribution of means, Single Sample and known population mean and standard deviation
9. B. if $t_{obt} \geq t_{crit}$
10. E. df
11. D. approaches the normal distribution
12. D. a & b
13. A. t test
14. D. the t distribution is a family of curves
15. B. n-1
16. Single sample t-test, t obtained = -2.07, t critical = -2.423, FAIL TO REJECT null. There was no significant reduction in weight (or there was an increase).

17a. standard error = 1.0041, z obtained = 4.48, z obt > z crit REJECT Null Hypothesis. Note: Another way to come to the same solution is to try and look up a z obtained of 4.48, the chart doesn't go that high but given as high as it does go, then $p < 0.0001$).

17b. standard error = 0.6755, z = 1.0067, Table A, Col C, $p = 0.33$ *Fail to Reject* Null hypothesis

18. N=45 (normal distribution), interval or better data, single sample, known population mean, unknown population standard deviation: single sample t test.

19. N=35 (normal distribution), single sample mean, known parameters of null hypothesis population: single sample z-test.

Chapter Seven Homework Answers

1. These designs reduce the effects of individual differences among subjects (which you are not likely to be studying). Reducing variability serves to increase the power of the test.
2. The sampling distribution of difference scores must be normal, and the dependent variable must be on the interval or ratio scale.
3. The sampling distribution of means must be normal, the samples must be drawn from populations having equal variances, and the dependent variable must be on the interval or ratio scale.
4. You could know that the population from which the scores come from is normally distributed, you could use the central limit theorem and assume that samples ≥ 30 are normally distributed, you could check that the skewness and kurtosis values for your data are not significantly different from normal and if they are different than normal, you could try to log transform them.
5. With the paired t-test, the subjects are the same in each group or they are "matched" on the characteristics that might affect the experiment. Thus, we can assume that both samples come from the same population and any differences we see are due to our independent variable. However, the independent t-test using random sampling to assign subjects to each group and we need to assume that both groups came from the same null hypothesis population with the same variance prior to our study.
6. Drop the subjects for which you have missing data points or do an independent t-test on all of your data points.
7. The paired t-test can be replaced with a non-parametric test called the Wilcoxon Signed Ranks Test while the independent t-test can be replaced with a non-parametric test called the Mann-Whitney U test. Another option is to perform a log transformation on your data to determine if the mathematical transformation will result in a normal distribution.
8. Regardless of the shape of the population of raw scores, the sampling distribution of the mean approaches a normal distribution as sample size N increases.
9. Independent t-test where t obt = -2.062, t crit = ± 2.306. So, Fail-to-Reject Null, being raised in a military or non-military family does not affect

the number of happy childhood memories you can recall.

10. Paired t-test where t obt = 3.13, t crit = $\pm$2.365. So, Reject the Null hypothesis. Race affects whether or not you are likely to shoot an innocent bystander.
11. Independent t-test where t obt = 1.542, t crit = 2.110. So, Fail-to-Reject Null, the day of the week does not appear to affect performance on exams.
12. n_1=31 and n_2=35 (normal distribution), two independent groups, interval or better data, s_1=1.34 and s_2=2.15 [$(2.15)^2/(1.34)^2 < 4$], so we have homogenous variances: <u>independent t test</u>
13. Two normally distributed paired samples, interval or better data: <u>paired t test</u>
14. N=45 (normal distribution), interval or better data, single sample, known population mean, unknown population standard deviation: <u>single sample t test</u>
15. N=35 (normal distribution), single sample mean, known parameters of null hypothesis population: <u>single sample z-test</u>

Chapter Eight Homework Answers

1. The variability among means increases relative to the variability within groups.

2. True, because F is a ratio of variance estimates (s_B^2/s_W^2), and since variances are always squared numbers, they are always positive, and ratio of two positive numbers is always positive.

3. For unplanned or post hoc comparisons, this test maintains your alpha level by adjusting the probability (critical value) for the number of means being compared (r). The number of means being compared is determined by ordering all of the means from smallest to largest, and then determining for each comparison whether there are means that fall between the two target means.

4. If there is no effect, then F-obtained will equal 1.00 (approximately). The reason F-obtained will equal 1.00 when there is no treatment effect is because you would be dividing one estimate of population variance by a second estimate of the same population variance, and any number divided by itself is equal to 1.00. Of course, F-obtained will rarely equal exactly 1.00 since our measurements of each estimate will vary due to random effects, but F-obtained values near 1.00 will never be significant because they will be suggesting little or no effect of our experimental treatment.

5. Both distributions form a family of curves based on degrees of freedom.

6. Use an independent ANOVA when you have more than two samples and a between-groups design, and when the sampling distribution is normally distributed, and when the dependent variable is on an interval or ratio scale, and when the variances of the groups are homogeneous.

7. Four-fold rule.

8.

Source	SumsSquares	df	(MS) s^2	F_{obt}	F_{crit}
Between	148.17	2	74.085	10.2976	4.26*
Within	64.75	9	7.1944		
Total	212.92		11		

* = Reject Null hypothesis!

Qobt =0.3728   Qobt = 4.2875*   Qobt = 4.6603*
Group C differs from A & B

9.

Source	SumsSquares	df	(MS) s^2	F_{obt}	F_{crit}
Between	3.2661	3	1.0887	2.32	2.72
Within	37.24	80	0.47		
Total	40.5061	83			

Fail-to-Reject Null Hypothesis!

10. One-way ANOVA because there are three independent groups, one variable, random sampling, normal distribution (all groups N is greater than or equal to 30), interval data (pitch discrimination is assessed based on the students' ability to tell if a note is separated by 7 or 4 half steps), homogeneous variance.

11. Single sample t-test because there is a single sample design in which the t assumptions of normality (16 subjects from a normally distributed population), random sampling, and interval or better data are met, but we do not know the population standard deviation.

12. Single sample z-test because we have both the population mean and standard deviation, a sample mean, random sampling of a normally distributed population (n=30)

13. Independent t test because there are two independent groups in which the independent t test assumptions of normality (n=35 for both groups), interval or better data (number of problems correct), random sampling, and homogeneous variances are met.

Chapter Nine Homework Answers

Provide a short answer for the following questions.

1. A repeated measures ANOVA is used when you have a within-groups design while a one-way ANOVA is used when you have a between-groups (independent) design.

2. Use a two-way ANOVA when you have more than two independent variables and a between-groups design, when the sampling distribution is normally distributed, the dependent variable is on an interval or ratio scale (although some statisticians believe that this is too strict), and when the variances of the groups are the same, or are homogeneous.

3. Use a repeated-measures ANOVA when you have one independent variable with more than two levels and a within-groups design, when the sampling distribution is normally distributed, and when the dependent variable is on an interval or ratio scale (although some statisticians believe that this is too strict).

4. They are the same EXCEPT that you have TWO independent variables with the two-way ANOVA which means that you calculate THREE F-obtained values (two main effects and one interaction).

5. It is an equation that is presented in the chapter to aid the conceptual understanding about the multiple factors that can influence an individual's score (including real effects of the IV's that we measure as well as unexplained variance).

6. Because you are able to determine how much of the previously unexplained variance (or error) can be explained by individual differences. Having multiple measures on the same individual allows us to calculate this value. Reducing the unexplained variance means

that we are able to reduce the size of our denominator (s_w^2), thus resulting in a larger F-obtained value (and increased power to detect any real effects of our experimental treatment).

7. A "main effect" is the influence of one of your independent variables on your dependent variable.

8. When the effect of one of your independent variables (or factors) is not the same at all levels of your other independent variable (or factor), then you have a statistical "interaction" effect.

9. Interaction terms occur when you have more than one independent variable. In the case of the repeated-measures ANOVA presented in this chapter, you only have one independent variable. The Subj variable is actually a subset of the old "within-groups variance" or error, and it would not be appropriate or interesting to determine interactions between "error" and your independent variable.

10.

ANOVA

Source of Variation	*SS*	*df*	*MS*	*F*	*F crit*
Row (sex)	828.8167	1	828.8167	3.030429	4.01954
Columns (treatment)	1265.633	2	632.8167	2.313788	3.168246
Interaction (sex * treatment)	847.2333	2	423.6167	1.548883	3.168246
Within	14768.9	54	273.4981		
Total	17710.58	59			

You must fail-to-reject the null hypotheses for the row, column and interaction values. There are no significant effects of your treatment or for sex.

11.

ANOVA

Source of Variation	SS	df	MS	F	F crit
Row (sex)	495.0625	1	495.0625	1.797164	4.012975
Columns (status)	1791.625	3	597.2083	2.167971	2.769433
Interaction (sex * status)	1855.063	3	618.3542	2.244734	2.769433
Within	15426.25	56	275.4688		
Total	19568	63			

You must fail-to-reject the null hypotheses for the row, column and interaction values. There are no significant effects of your treatment or for sex.

12.

ANOVA

Source of Variation	SS	df	MS	F	F crit
Row (job)	560952.5	1	560952.5	2.242676	4.08474
Columns (blood pressure)	10028590	3	3342863	13.36469	2.838746
Interaction (job * blood pressure)	199493.2	3	66497.74	0.265857	2.838746
Within	10005058	40	250126.4		
Total	20794093	47			

You must fail-to-reject the null hypotheses for the row and interaction values. There are no significant effects for job or for the interaction between job and blood pressure. However, there are significant differences between your blood pressure groups.

13. One-way ANOVA. One IV (school) with five levels, and a between-groups design. Assumptions are met (interval or ratio test scores, sample sizes large enough to assume normality, homogeneity of variance was assumed in the instructions).

14. Independent t-test. Only two groups of subjects in a between groups design. The same size is large enough that normality can be assumed

and the variances between the groups are homogenous. The number of details is an interval or ratio scale of measurement.

15. Repeated measures ANOVA. One group of 35 individuals was eventually analyzed for four time periods. So you had four scores for each person. The sample size is sufficient to assume normality. The homogeneity of variance assumption was thrown in as a distracter. It doesn't matter if the variances are homogeneous for this test. The memory scores were interval or ratio data.

Chapter Ten Homework Answers

1. It is a measure of how much variance in variable Y can be accounted for by variable X. It is calculated by squaring the value of the correlation between X and Y (r^2).

2. $r^2 = (0.89)^2 = 0.7921 = 0.79$

3. It tells us that 79% of the variability found in Y can be accounted for by X.
4. From –1.00 to 1.00

5. The shape of the relationship between X and Y, and the measurement scale underlying the data

6. That the relationship between X and Y is linear.

7.
$\Sigma X =$ 19
$\Sigma X^2 =$ 71
$\Sigma XY =$ 1,782
$\Sigma Y =$ 1,035
$\Sigma Y^2 =$ 108,293
r = **-0.91**

8. $Y' = b_y X + a_y$; $Y' = -5.286533(X) + 113.54441$

9.(c) both a and b are true

10. (b) a relationship exists, but all of the points do not fall on the line

11. (a) higher the correlation

12. (d) 0.15

13. (e) none of the above

14. (b) they should fire that sorry statistician and hire a new one (of course: an answer of "-1.08" is impossible!)

15a. -0.52
15b. –0.04
15c. 100.61
15d. 85.33
15e. The standard error of the estimate (SEE) = 8.22 and is quite large. The R^2 is approximately 0.27, so reaction time only explains 27% of the variability in vocabulary scores. I wouldn't be very confident about the prediction.

16. One-way ANOVA because there are three independent groups, one variable, random sampling, normal distribution (all groups N is greater than or equal to 30), interval data homogeneous variance.

17. Single sample t-test because there is a single sample design in which the t assumptions of normality (16 subjects from a normally distributed population), random sampling, and interval or better data are met, but we do not know the population standard deviation.

18. Single sample z-test because we have both the population mean and standard deviation, a sample mean, random sampling of a normally distributed population (n=30).

19. Independent t test because there are two independent groups in which the independent t test assumptions of normality (n=35 for both groups), interval or better data (number of problems correct), random sampling, and homogeneous variances are met.

20. Paired t test because we have before and after conditions with the same subjects participating in both conditions. We have a normal distribution because our n was greater that 30 and the data were measured on a ratio scale.

Chapter Eleven Homework Answers

1. You can understand and interpret why there is an F-obtained value in your linear regression output. You can mix and match the measurement scales of your independent variables (this also allows you to analyze interactions amongst ordinal/nominal and interval/ratio variables).

2. Because you are only analyzing one dependent variable and the test is named for the number of dependent variables. An analysis with two dependent variables would be a multivariate test.

3. Analysis of Covariance

4. ANCOVA, is a specialized form of ANOVA that is used when an investigator wishes to remove the effects of a variable that is known to influence the dependent variable, but is not the subject of the current experiment and analysis.

5. They are all based on a general linear equation where there are multiple influences on any given score. The effects of the grand mean, the independent variable and the unexplained variance (error) are all included in each conceptual formula.

6. ANOVA, regression and t-tests were developed independently of one another, and thus their terminology varies historically. The realization that all of these techniques are really the same and have the same underlying linear (additive) equations is more recent, so the traditional terminology associated with each test independent of the other tests is still widely prevalent in the literature.

Chapter Twelve Homework Answers

1. You are likely to reject the null hypothesis.

2. The scores in each cell are independent and no f_e is less than 5 if the number of rows or columns is greater than 2.

3. The dependent variable or the difference scores must be ordinal.

4. The dependent variable must be ordinal or better.

5. The dependent variable must be ordinal or better and if you are using the Chi-Square table, then there must be at least five scores in each sample.

6. Used when one or both of the variables are ordinal and the relationship between the variables is linear.

7. You do not use the Chi-square test if fe is below 5. If the experiment involves a 2 by 2 contingency table and the fe is below 5, then a Fisher's exact probability test can be used.

8. A contingency table is a two-way chi square. In other words, there are two factors with multiple levels rather than one factor, and you are testing the likelihood that the frequencies are randomly distributed amongst the cells in the table.

9. Independent t –test (Mann-Whitney U test)
   Paired t-test (Wilcoxon)
   ANOVA (Kruskal-Wallis)

10. a. frequencies

11. $\chi^2_{obtained} = 2.0906$, df= 1, $\chi^2_{critical} = 3.841$, Fail-to-reject null hypothesis. Age and soda preference appear to be unrelated.

12. $\chi^2_{obtained} = 38.0953$, df= 2, $\chi^2_{critical} = 5.991$, Reject null hypothesis. Gender and watch preference appear to be related.

13. $\chi^2_{obtained} = 8.0000$, df= 1, $\chi^2_{critical} = 6.635$, Reject null hypothesis. The distribution of opinions is not random.

14. $\chi^2_{obtained} = 53.662491$, $\chi^2_{critical} = 5.991$, Reject null hypothesis. The distribution of opinions is not random. Conclude that reading performance has changed.

15. Kruskal-Wallis test because there are three independent groups, one variable and random sampling, but the ANOVA assumptions are not met. The data is not normally distributed because n is not greater than or equal to 30 (n=3) and we were not told to assume normality.

16. Wilcoxon-signed ranks test because we have a paired design in which the t test assumptions are not met. Although the data is interval or better (heartbeats per minute), we do not have a normal distribution (n is not greater than or equal to 30). The Wilcoxon-signed ranks test was chosen over the sign test because it is more powerful.

17. Mann-Whitney U test because it is an independent groups design with two conditions and doesn't meet the assumptions for the independent t test. Although the ratio scaled data is randomly selected from a normally distributed population, the variances are not homogeneous.

18. Wilcoxon signed ranks test because there are two correlated groups in which the subject's anxiety after treatment is compared to subject's baseline anxiety. The assumptions of normality and random sampling are met. The measurement, however, is ordinal.

Chapter Thirteen Homework Answers

1.Mann-Whitney U test because the homogeneity of variance assumption is not met.

2.Wilcoxon Signed Ranks test because you cannot assume normality and the measurement scale is ordinal.

3. Independent t-test because you have two groups and the assumptions are met.

4. Kruskal-Wallis test because there are four wines, but the ratings are ordinal.

5. Single Sample z-test because you know the population mean and standard deviation.

6. Paired t-test because the design is paired/correlated and you can assume normality based on the sample size.

7. Contingency Table Chi-Square because frequencies are the dependent variable and there are two factors.

8. ANOVA because you have three groups and the assumptions are met.

9. Mann-Whitney U-test because you cannot assume normality.

10. Paired t-test because the subjects are matched individually with similar players and you can assume normality based on sample size.

11. Contingency Table Chi-Square because you have two factors and the dependent variable is frequencies.

12. Chi-Square because the data are frequencies (nominal) and you have one factor.

13. ANOVA because the assumptions are met and you have three groups.

14. Paired t-test because you can compare lemon preference before and after conditioning on a large sample size, so you can assume normality.

15. Wilcoxon Signed Rank test because the sample size is unknown and you cannot assume normality.

Statistical Tables

Areas under the normal distribution curve

z A	Area Between Mean and z B	Area Beyond z C	z A	Area Between Mean and z B	Area Beyond z C
0.00	0.0000	0.5000	0.31	0.1217	0.3783
0.01	0.0040	0.4960	0.32	0.1255	0.3745
0.02	0.0080	0.4920	0.33	0.1293	0.3707
0.03	0.0120	0.4880	0.34	0.1331	0.3669
0.04	0.0160	0.4840	0.35	0.1368	0.3632
0.05	0.0199	0.4801	0.36	0.1406	0.3594
0.06	0.0239	0.4761	0.37	0.1443	0.3557
0.07	0.0279	0.4721	0.38	0.1480	0.3520
0.08	0.0319	0.4681	0.39	0.1517	0.3483
0.09	0.0359	0.4641	0.40	0.1554	0.3446
0.10	0.0398	0.4602	0.41	0.1591	0.3409
0.11	0.0438	0.4562	0.42	0.1628	0.3372
0.12	0.0478	0.4522	0.43	0.1664	0.3336
0.13	0.0517	0.4483	0.44	0.1700	0.3300
0.14	0.0557	0.4443	0.45	0.1736	0.3264
0.15	0.0596	0.4404	0.46	0.1772	0.3228
0.16	0.0636	0.4364	0.47	0.1808	0.3192
0.17	0.0675	0.4325	0.48	0.1844	0.3156
0.18	0.0714	0.4286	0.49	0.1879	0.3121
0.19	0.0753	0.4247	0.50	0.1915	0.3085
0.20	0.0793	0.4207	0.51	0.1950	0.3050
0.21	0.0832	0.4168	0.52	0.1985	0.3015
0.22	0.0871	0.4129	0.53	0.2019	0.2981
0.23	0.0910	0.4090	0.54	0.2054	0.2946
0.24	0.0948	0.4052	0.55	0.2088	0.2912
0.25	0.0987	0.4013	0.56	0.2123	0.2877
0.26	0.1026	0.3974	0.57	0.2157	0.2843
0.27	0.1064	0.3936	0.58	0.2190	0.2810
0.28	0.1103	0.3897	0.59	0.2224	0.2776
0.29	0.1141	0.3859	0.60	0.2257	0.2743
0.30	0.1179	0.3821	0.61	0.2291	0.2709

Areas under the normal distribution curve, contd.

z A	Area Between Mean and z B	Area Beyond z C	z A	Area Between Mean and z B	Area Beyond z C
0.62	0.2324	0.2676	0.91	0.3186	0.1814
0.63	0.2357	0.2643	0.92	0.3212	0.1788
0.64	0.2389	0.2611	0.93	0.3238	0.1762
0.65	0.2422	0.2578	0.94	0.3264	0.1736
0.66	0.2454	0.2546	0.95	0.3289	0.1711
0.67	0.2486	0.2514	0.96	0.3315	0.1685
0.68	0.2517	0.2483	0.97	0.3340	0.1660
0.69	0.2549	0.2451	0.98	0.3365	0.1635
0.70	0.2580	0.2420	0.99	0.3389	0.1611
0.71	0.2611	0.2389	1.00	0.3413	0.1587
0.72	0.2642	0.2358	1.01	0.3438	0.1562
0.73	0.2673	0.2327	1.02	0.3461	0.1539
0.74	0.2704	0.2296	1.03	0.3485	0.1515
0.75	0.2734	0.2266	1.04	0.3508	0.1492
0.76	0.2764	0.2236	1.05	0.3531	0.1469
0.77	0.2794	0.2206	1.06	0.3554	0.1446
0.78	0.2823	0.2177	1.07	0.3577	0.1423
0.79	0.2852	0.2148	1.08	0.3599	0.1401
0.80	0.2881	0.2119	1.09	0.3621	0.1379
0.81	0.2910	0.2090	1.10	0.3643	0.1357
0.82	0.2939	0.2061	1.11	0.3665	0.1335
0.83	0.2967	0.2033	1.12	0.3686	0.1314
0.84	0.2995	0.2005	1.13	0.3708	0.1292
0.85	0.3023	0.1977	1.14	0.3729	0.1271
0.86	0.3051	0.1949	1.15	0.3749	0.1251
0.87	0.3078	0.1922	1.16	0.3770	0.1230
0.88	0.3106	0.1894	1.17	0.3790	0.1210
0.89	0.3133	0.1867	1.18	0.3810	0.1190
0.90	0.3159	0.1841	1.19	0.3830	0.1170

Areas under the normal distribution curve, contd.

z A	Area Between Mean and z B	Area Beyond z C	z A	Area Between Mean and z B	Area Beyond z C
1.20	0.3849	0.1151	1.50	0.4332	0.0668
1.21	0.3869	0.1131	1.51	0.4345	0.0655
1.22	0.3888	0.1112	1.52	0.4357	0.0643
1.23	0.3907	0.1093	1.53	0.4370	0.0630
1.24	0.3925	0.1075	1.54	0.4382	0.0618
1.25	0.3944	0.1056	1.55	0.4394	0.0606
1.26	0.3962	0.1038	1.56	0.4406	0.0594
1.27	0.3980	0.1020	1.57	0.4418	0.0582
1.28	0.3997	0.1003	1.58	0.4429	0.0571
1.29	0.4015	0.0985	1.59	0.4441	0.0559
1.30	0.4032	0.0968	1.60	0.4452	0.0548
1.31	0.4049	0.0951	1.61	0.4463	0.0537
1.32	0.4066	0.0934	1.62	0.4474	0.0526
1.33	0.4082	0.0918	1.63	0.4484	0.0516
1.34	0.4099	0.0901	1.64	0.4495	0.0505
1.35	0.4115	0.0885	1.65	0.4505	0.0495
1.36	0.4131	0.0869	1.66	0.4515	0.0485
1.37	0.4147	0.0853	1.67	0.4525	0.0475
1.38	0.4162	0.0838	1.68	0.4535	0.0465
1.39	0.4177	0.0823	1.69	0.4545	0.0455
1.40	0.4192	0.0808	1.70	0.4554	0.0446
1.41	0.4207	0.0793	1.71	0.4564	0.0436
1.42	0.4222	0.0778	1.72	0.4573	0.0427
1.43	0.4236	0.0764	1.73	0.4582	0.0418
1.44	0.4251	0.0749	1.74	0.4591	0.0409
1.45	0.4265	0.0735	1.75	0.4599	0.0401
1.46	0.4279	0.0721	1.76	0.4608	0.0392
1.47	0.4292	0.0708	1.77	0.4616	0.0384
1.48	0.4306	0.0694	1.78	0.4625	0.0375
1.49	0.4319	0.0681	1.79	0.4633	0.0367

Areas under the normal distribution curve, contd.

z A	Area Between Mean and z B	Area Beyond z C	z A	Area Between Mean and z B	Area Beyond z C
1.80	0.4641	0.0359	2.10	0.4821	0.0179
1.81	0.4649	0.0351	2.11	0.4826	0.0174
1.82	0.4656	0.0344	2.12	0.4830	0.0170
1.83	0.4664	0.0336	2.13	0.4834	0.0166
1.84	0.4671	0.0329	2.14	0.4838	0.0162
1.85	0.4678	0.0322	2.15	0.4842	0.0158
1.86	0.4686	0.0314	2.16	0.4846	0.0154
1.87	0.4693	0.0307	2.17	0.4850	0.0150
1.88	0.4699	0.0301	2.18	0.4854	0.0146
1.89	0.4706	0.0294	2.19	0.4857	0.0143
1.90	0.4713	0.0287	2.20	0.4861	0.0139
1.91	0.4719	0.0281	2.21	0.4864	0.0136
1.92	0.4726	0.0274	2.22	0.4868	0.0132
1.93	0.4732	0.0268	2.23	0.4871	0.0129
1.94	0.4738	0.0262	2.24	0.4875	0.0125
1.95	0.4744	0.0256	2.25	0.4878	0.0122
1.96	0.4750	0.0250	2.26	0.4881	0.0119
1.97	0.4756	0.0244	2.27	0.4884	0.0116
1.98	0.4761	0.0239	2.28	0.4887	0.0113
1.99	0.4767	0.0233	2.29	0.4890	0.0110
2.00	0.4772	0.0228	2.30	0.4893	0.0107
2.01	0.4778	0.0222	2.31	0.4896	0.0104
2.02	0.4783	0.0217	2.32	0.4898	0.0102
2.03	0.4788	0.0212	2.33	0.4901	0.0099
2.04	0.4793	0.0207	2.34	0.4904	0.0096
2.05	0.4798	0.0202	2.35	0.4906	0.0094
2.06	0.4803	0.0197	2.36	0.4909	0.0091
2.07	0.4808	0.0192	2.37	0.4911	0.0089
2.08	0.4812	0.0188	2.38	0.4913	0.0087
2.09	0.4817	0.0183	2.39	0.4916	0.0084

Areas under the normal distribution curve, contd.

z A	Area Between Mean and z B	Area Beyond z C	z A	Area Between Mean and z B	Area Beyond z C
2.40	0.4918	0.0082	2.70	0.4965	0.0035
2.41	0.4920	0.0080	2.71	0.4966	0.0034
2.42	0.4922	0.0078	2.72	0.4967	0.0033
2.43	0.4925	0.0075	2.73	0.4968	0.0032
2.44	0.4927	0.0073	2.74	0.4969	0.0031
2.45	0.4929	0.0071	2.75	0.4970	0.0030
2.46	0.4931	0.0069	2.76	0.4971	0.0029
2.47	0.4932	0.0068	2.77	0.4972	0.0028
2.48	0.4934	0.0066	2.78	0.4973	0.0027
2.49	0.4936	0.0064	2.79	0.4974	0.0026
2.50	0.4938	0.0062	2.80	0.4974	0.0026
2.51	0.4940	0.0060	2.81	0.4975	0.0025
2.52	0.4941	0.0059	2.82	0.4976	0.0024
2.53	0.4943	0.0057	2.83	0.4977	0.0023
2.54	0.4945	0.0055	2.84	0.4977	0.0023
2.55	0.4946	0.0054	2.85	0.4978	0.0022
2.56	0.4948	0.0052	2.86	0.4979	0.0021
2.57	0.4949	0.0051	2.87	0.4979	0.0021
2.58	0.4951	0.0049	2.88	0.4980	0.0020
2.59	0.4952	0.0048	2.89	0.4981	0.0019
2.60	0.4953	0.0047	2.90	0.4981	0.0019
2.61	0.4955	0.0045	2.91	0.4982	0.0018
2.62	0.4956	0.0044	2.92	0.4982	0.0018
2.63	0.4957	0.0043	2.93	0.4983	0.0017
2.64	0.4959	0.0041	2.94	0.4984	0.0016
2.65	0.4960	0.0040	2.95	0.4984	0.0016
2.66	0.4961	0.0039	2.96	0.4985	0.0015
2.67	0.4962	0.0038	2.97	0.4985	0.0015
2.68	0.4963	0.0037	2.98	0.4986	0.0014
2.69	0.4964	0.0036	2.99	0.4986	0.0014

Areas under the normal distribution curve, contd.

z A	Area Between Mean and z B	Area Beyond z C	z A	Area Between Mean and z B	Area Beyond z C
3.00	0.4987	0.0013	3.30	0.4995	0.0005
3.01	0.4987	0.0013	3.31	0.4995	0.0005
3.02	0.4987	0.0013	3.32	0.4995	0.0005
3.03	0.4988	0.0012	3.33	0.4996	0.0004
3.04	0.4988	0.0012	3.34	0.4996	0.0004
3.05	0.4989	0.0011	3.35	0.4996	0.0004
3.06	0.4989	0.0011	3.36	0.4996	0.0004
3.07	0.4989	0.0011	3.37	0.4996	0.0004
3.08	0.4990	0.0010	3.38	0.4996	0.0004
3.09	0.4990	0.0010	3.39	0.4997	0.0003
3.10	0.4990	0.0010	3.40	0.4997	0.0003
3.11	0.4991	0.0009	3.41	0.4997	0.0003
3.12	0.4991	0.0009	3.42	0.4997	0.0003
3.13	0.4991	0.0009	3.43	0.4997	0.0003
3.14	0.4992	0.0008	3.44	0.4997	0.0003
3.15	0.4992	0.0008	3.45	0.4997	0.0003
3.16	0.4992	0.0008	3.46	0.4997	0.0003
3.17	0.4992	0.0008	3.47	0.4997	0.0003
3.18	0.4993	0.0007	3.48	0.4997	0.0003
3.19	0.4993	0.0007	3.49	0.4998	0.0002
3.20	0.4993	0.0007	3.50	0.4998	0.0002
3.21	0.4993	0.0007	3.51	0.4998	0.0002
3.22	0.4994	0.0006	3.52	0.4998	0.0002
3.23	0.4994	0.0006	3.53	0.4998	0.0002
3.24	0.4994	0.0006	3.54	0.4998	0.0002
3.25	0.4994	0.0006	3.55	0.4998	0.0002
3.26	0.4994	0.0006	3.56	0.4998	0.0002
3.27	0.4995	0.0005	3.57	0.4998	0.0002
3.28	0.4995	0.0005	3.58	0.4998	0.0002
3.29	0.4995	0.0005	3.59	0.4998	0.0002

Areas under the normal distribution curve, contd.

z A	Area Between Mean and z B	Area Beyond z C
3.60	0.4998	0.0002
3.61	0.4998	0.0002
3.62	0.4999	0.0001
3.63	0.4999	0.0001
3.64	0.4999	0.0001
3.65	0.4999	0.0001
3.66	0.4999	0.0001
3.67	0.4999	0.0001
3.68	0.4999	0.0001
3.69	0.4999	0.0001
3.70	0.4999	0.0001

Critical values of the Student t distribution

	Level of Significance for One-tailed Test					
	0.10	0.05	0.025	0.010	0.005	0.0005
	Level of Significance for Two-tailed Test					
df	0.20	0.10	0.05	0.02	0.01	0.001
1	3.078	6.314	12.706	31.821	63.657	636.619
2	1.886	2.920	4.303	6.965	9.925	31.599
3	1.638	2.353	3.182	4.541	5.841	12.924
4	1.533	2.132	2.776	3.747	4.604	8.610
5	1.476	2.015	2.571	3.365	4.032	6.869
6	1.440	1.943	2.447	3.143	3.707	5.959
7	1.415	1.895	2.365	2.998	3.499	5.408
8	1.397	1.860	2.306	2.896	3.355	5.041
9	1.383	1.833	2.262	2.821	3.250	4.781
10	1.372	1.812	2.228	2.764	3.169	4.587
11	1.363	1.796	2.201	2.718	3.106	4.437
12	1.356	1.782	2.179	2.681	3.055	4.318
13	1.350	1.771	2.160	2.650	3.012	4.221
14	1.345	1.761	2.145	2.624	2.977	4.140
15	1.341	1.753	2.131	2.602	2.947	4.073
16	1.337	1.746	2.120	2.583	2.921	4.015
17	1.333	1.740	2.110	2.567	2.898	3.965
18	1.330	1.734	2.101	2.552	2.878	3.922
19	1.328	1.729	2.093	2.539	2.861	3.883
20	1.325	1.725	2.086	2.528	2.845	3.850
21	1.323	1.721	2.080	2.518	2.831	3.819
22	1.321	1.717	2.074	2.508	2.819	3.792
23	1.319	1.714	2.069	2.500	2.807	3.768
24	1.318	1.711	2.064	2.492	2.797	3.745
25	1.316	1.708	2.060	2.485	2.787	3.725
26	1.315	1.706	2.056	2.479	2.779	3.707
27	1.314	1.703	2.052	2.473	2.771	3.690
28	1.313	1.701	2.048	2.467	2.763	3.674
29	1.311	1.699	2.045	2.462	2.756	3.659
30	1.310	1.697	2.042	2.457	2.750	3.646
40	1.303	1.684	2.021	2.423	2.704	3.551
60	1.296	1.671	2.000	2.390	2.660	3.460
120	1.289	1.658	1.980	2.358	2.617	3.373
∞	1.282	1.645	1.960	2.327	2.576	3.291